Contents

This edition published exclusively for WHSmith, 1998, by
Stanley Thornes (Publishers) Ltd
Ellenborough House
Wellington Street
CHELTENHAM
GL50 1YW

98 99 00 01 02 \ 10 9 8 7 6 5 4 3 2 1

A catalogue record for this book is available from the British Library

ISBN 0-7487-3581-X

Designed and typeset by Ian Foulis & Associates, Saltash, Cornwall
Illustrated by Ian Foulis & Associates

Printed and bound in Spain by Mateu Cromo

Introduction

What this book contains

During your child's third year in secondary school (Year 9) he or she will sit Key Stage 3 National Assessment Tests in the three core subjects: English, mathematics and science. These tests take place in school over a period of about a week during May and results are reported back to you. For each of the three subjects your child will be given a mark in the form of a level. Most children will perform in the range of Levels 4-6 by the end of Key Stage 3, with an average performance being roughly Level 5.

The tests are a valuable measure of your child's performance over the first three years of secondary school and give a likely indication of their likely performance at Key Stage 4, which leads to the GCSE examinations.

This book provides practice papers and questions in mathematics to be used to prepare your child confidently for the tests in this subject. The papers will:

- provide test questions similar to those in the National Tests for Key Stage 3 of the National Curriculum;
- give your child practice in sitting the tests: working to a set time, getting familiar with the format and style of the tests and developing effective test strategies;
- give a broad guide to your child's likely level of performance;
- give you an idea of strengths and weaknesses in your child's learning.

What the National Tests in mathematics cover

The National Curriculum for mathematics at Key Stage 3 is divided into four areas called 'attainment targets'.

The four attainment targets are:

1. Using and Applying Mathematics
2. Number and Algebra
3. Shape, Space and Measures
4. Handling Data

The first attainment target, Using and Applying Mathematics, involves practical and investigational work relating to the other three attainment targets. It does not form part of the National Tests but is assessed by the school.

For the tests in May your child will probably sit three papers, two of these being ordinary written papers (one of which will be a 'no calculator' paper) and the third being a mental test. There are four tiers of entry: National Curriculum Levels 3-5, 4-6, 5-7 and 6-8.

What levels do these practice papers cover?

For ease of use and also to give a student maximum practice, this book contains two written tests and two mental tests. These cover the most popular tiers of entry at Levels 4-6 and 5-7. Separate practice questions at Level 3 and Level 8 are also provided. Each paper contains 15 questions, five questions at each of the Levels covered. Paper 1 has questions 1-5 at Level 4, 6-10 at Level 5 and 11-15 at Level 6. Paper 2 has 1-5 at Level 5, 6-10 at Level 6 and 11-15 at Level 7. Paper 1 is a 'no calculator' paper, but a calculator can be used for Paper 2. There are five practice questions at Level 3 and five at Level 8. Questions where a calculator is allowed are marked accordingly.

Mental Paper 1 is aimed at the expected level of attainment for pupils at Level 4.

Mental Paper 2 is aimed at the expected level of attainment for pupils at Level 5-6.

Helping your child sit tests

As well as practising mathematics, one of the key aims of this book is to give your child practice in working under test conditions. All the tests are timed and your child should try to complete each one within the given time. In order to make the best use of the tests, and to ensure that the experience is a positive one for your child, it is helpful to follow a few basic principles:

- Talk with your child first before beginning the tests. Present the activity positively and reassuringly. Encourage your child to consider the papers as an activity which will help, always making him or her feel secure about the process. Remember that an average student will not have studied all the topics included in these tests.
- Ensure that your child is relaxed and rested before doing a test. It may be better to do a paper at the weekend or during the holidays rather than straight after a day at school.
- Ensure a quiet place, free from noise or disturbance, for doing the tests.
- Ensure that there is a watch or clock available (with a second hand when doing the mental tests). Your child will also need:
  - a pencil and a ruler
  - a calculator, preferably a scientific calculator but there should be a √ key
  - a protractor or angle measurer
  - tracing paper
- Progress through the tests in the order of the levels. Be positive about what your child achieves. Remember that an average student at Key Stage 3 is only at Level 5 but these papers include material up to Level 8.
- Ensure that your child understands exactly what to do for each paper and give some basic strategies for tackling the task. For example:
  - Try to tackle all the questions but don't worry if you can't do some. Put a pencil mark by any you can't do, leave them and come back to them at the end.
  - Make sure you read the questions carefully.
  - Refer to the formula sheet on page 6.
  - If you have any time over at the end, go back and check your answers.
- Taking time to talk over a test beforehand and to discuss any difficulties afterwards will help your child to become more confident at sitting tests.
- However your child does, ensure that you give plenty of praise for effort.

Setting the mental tests

You can use the mental tests before or after the written papers. Your child will probably find them quite challenging if they have not attempted anything similar before. For this reason there are two answer sheets for each test. It is suggested that you start with Mental Test 1 to be answered on sheet 1A. A few days later try the same test again, giving your child sheet 1B for recording the answers. If your child coped fairly well with Mental Test 1, try Mental Test 2 in the same way. You will need to remove the answer sheets from the book to give to your child.

Before you begin, it would be helpful to practise the following skills with your child:

- tables up to 10×10;
- adding and subtracting mentally, for example: $18 + 7$ or $36 - 12$;
- strategies such as 'to get $80 - 48$, first do $80 - 50$, then take away 2';
- for Level 5, multiply or divide a number, including a decimal, by 10, 100 or 1 000; for example: $34 \div 10$ or 7.5×100;
- for Level 7, check calculations by rounding to one significant figure; for example: $3\,145 \div 5.7$ is about 3 000 divided by 6 which is 500.

You will need to ensure that you read the questions to your child within the set time.

Setting the written papers

Try starting with the Level 3 practice questions, which should take about 15 minutes. Allow your child 20 minutes maximum. If your child finds these relatively easy, then give them Paper 1. Allow the student one hour to complete this. If your child completes Paper 1 then let him or her try all or part of Paper 2. Allow one hour for Paper 2 as well. Only attempt the Level 8 questions if your child gets close to Level 7 on Paper 2.

Before you begin, it would also be useful to make sure that when working on paper your child can:

- at Level 4, add and subtract decimals and do short multiplication and division by whole numbers; for example, set out properly and work out $3.6 + 15.2$ or 5.04×7;
- at Level 5, do long multiplication or long division by whole two-digit numbers; for example, set out properly and work out 6.82×27 or $598 \div 23$.

What to do with the results

The tests in this book and the results gained from them are only a guide to your child's likely level of performance. They are not an absolute guarantee of how your child will perform in the National Tests themselves. However, these papers will at least allow your child to gain practice in sitting tests; they will also give you an insight into the strengths and weaknesses in their learning.

If there are particular areas of performance which seem weaker, it may be worth providing more practice of the skills required. It is also valuable to discuss any such weaknesses with your child's subject teacher, and to seek confirmation of any problem areas and advice on how to proceed. It is always better to work in partnership with the school if you can. Above all ensure that you discuss these issues with your child in a positive and supportive way so that you have their co-operation in working together to improve learning.

Formulae

You may need to use some of these formulae.

AREA

Circle

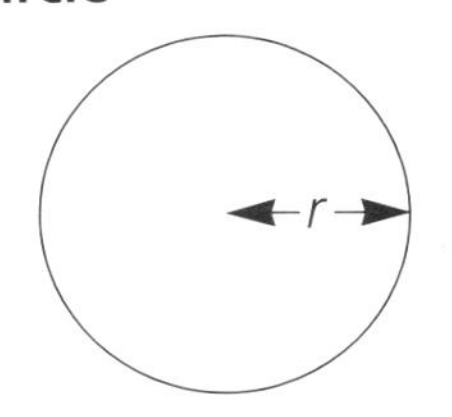

πr^2

Take π as 3.14 or use the π key on your calculator

Rectangle

length × width

Triangle

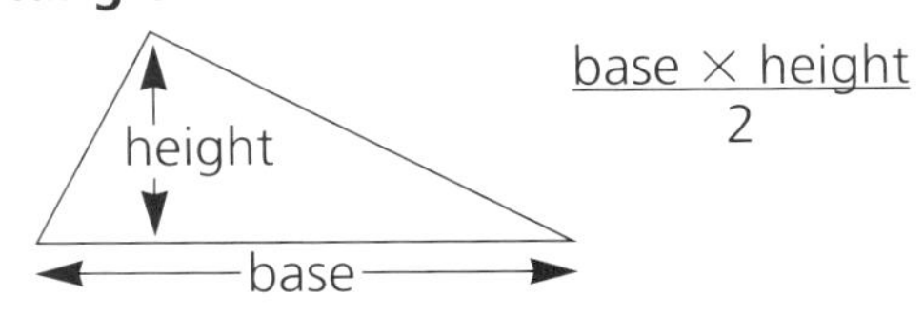

$\dfrac{\text{base} \times \text{height}}{2}$

Parallelogram

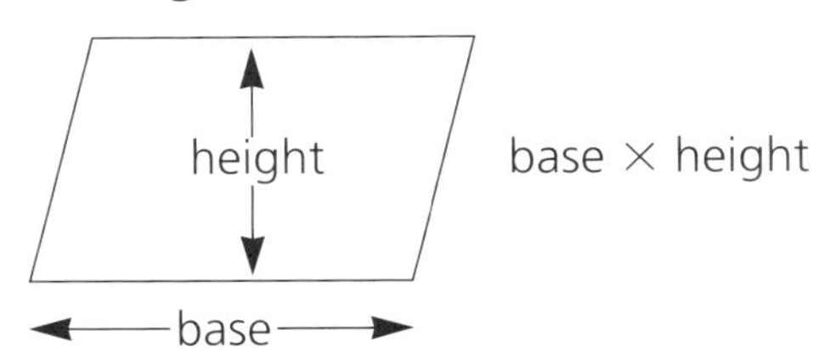

base × height

Trapezium

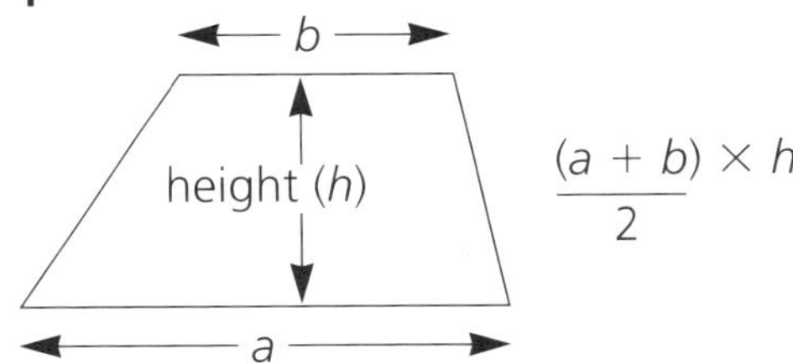

$\dfrac{(a + b) \times h}{2}$

LENGTH

Circle

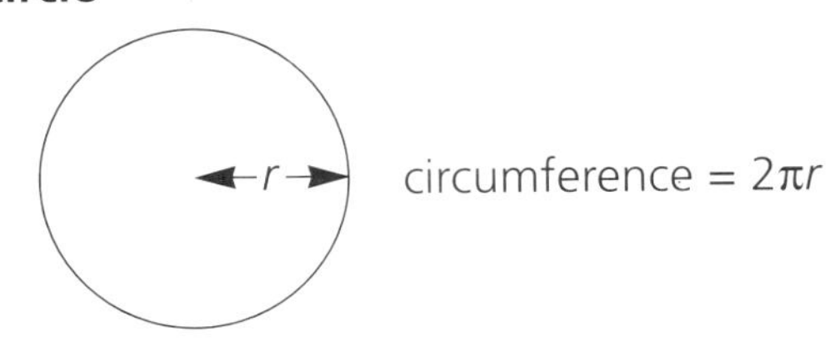

circumference = $2\pi r$

For a right-angled triangle

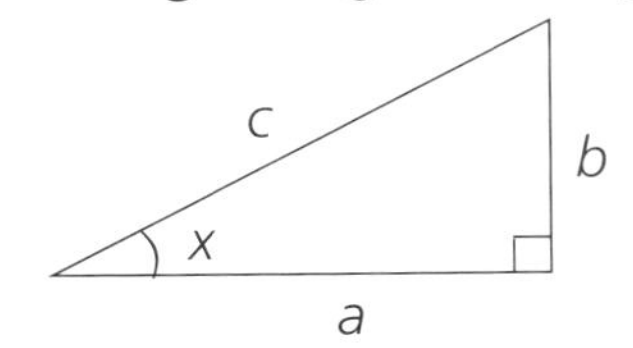

$a^2 + b^2 = c^2$ (Pythagoras' theorem)

$a = c \cos x$	$\cos x = \dfrac{a}{c}$
$b = c \sin x$	$\sin x = \dfrac{b}{c}$
$b = a \tan x$	$\tan x = \dfrac{b}{a}$

VOLUME

Prism

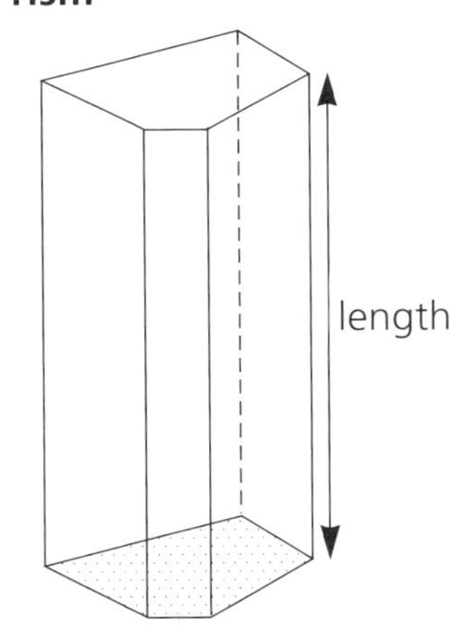

area of cross-section × length

Mental Test 1

You will need a watch with a second hand.

The student should *only* have the answer sheet and a pen or pencil; calculators, rulers, erasers, and so on, are not allowed.

Read each question to the student twice.

The student then has the stated number of seconds to write the answer. Before you begin, explain to your child the following:

- you must work out the answers in your head – calculators and written workings out are not allowed;
- you must record each answer on the answer sheet, in the box alongside that question number;
- if you make a mistake, cross out the wrong answer and write the new answer alongside the box;
- it is quicker to write answers as numbers rather than words;
- if you cannot answer a question put a line in the box;

Now begin the test. (Remember to repeat each question.)

You will have 5 seconds to answer each of these questions.

1 Write thirty thousand and twenty in figures.

2 Write the number that is four less than one hundred.

3 What is five hundred and seventy-four to the nearest hundred?

4 What is six multiplied by eight?

5 Write nought point three as a fraction.

6 What is eighty-five multiplied by ten?

7 What is thirty-two divided by four?

8 Write two and a half metres in centimetres.

You will have 10 seconds to answer each of these questions.

9 Add together seven, three and sixteen.

10 How many halves are there in five whole ones?

11 If forty-four per cent of the children in a school are boys, what percentage are girls?

12 The side of a square is five metres. What is the area of the square?

13 A bus journey starts at eight forty-five in the morning. It lasts for forty minutes. At what time does the bus arrive?

14 The temperature is minus three degrees Celsius. It rises by eight degrees. What is the new temperature?

15 On your sheet is part of a scale. Write down the exact number shown by the arrow.

16 The numbers one and twenty-one are factors of twenty-one. Write down two other factors of twenty-one.

17 What is six thousand divided by ten?

18 What number is four squared?

19 Sally has seven pounds fifty pence. She wants to buy a shirt that costs fourteen pounds thirty pence. How much more money does Sally need?

20 Fifty per cent of a number is seventeen. What is the number?

21 Look at the angle drawn on your sheet. Estimate its size in degrees.

You will have 15 seconds to answer each of these questions.

22 John wants to buy a book which costs three pounds twenty pence. He saves fifty pence a week. In how many weeks will he have enough money to buy the book?

23 Look at the numbers in the box on your answer sheet. Circle all of the odd numbers.

24 Add twenty-eight and forty-three.

25 Wallpaper costs one pound ninety-nine a roll. How much will five rolls cost?

26 Subtract eighteen from sixty-three.

27 Look at the two numbers on the answer sheet. Write the number that is halfway between them.

28 One quarter of a number is thirty-five. What is the number?

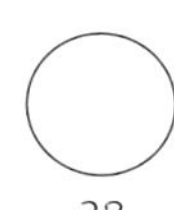

Mental Test 2

You will need a watch with a second hand.

The student should *only* have the answer sheet and a pen or pencil; calculators, rulers, erasers, and so on, are not allowed.

Read each question to the student twice.

The student then has the stated number of seconds to write the answer. Before you begin, explain to your child the following:

- you must work out the answers in your head – calculators and written workings out are not allowed;
- you must record each answer on the answer sheet, in the box alongside that question number;
- if you make a mistake, cross out the wrong answer and write the new answer alongside the box;
- it is quicker to write answers as numbers rather than words;
- if you cannot answer a question put a line in the box;

Now begin the test. (Remember to repeat each question.)

You will have 5 seconds to answer each of these questions.

1 What is one hundred and thirteen multiplied by ten?

2 Write one hundred and seventy millimetres in centimetres.

3 What is sixty-four divided by eight?

4 What is five point nine multiplied by one hundred?

5 Write seven hundredths as a decimal number.

6 A line is measured as twelve millimetres to the nearest millimetre. What is the minimum length the line could be?

You will have 10 seconds to answer each of these questions.

7 A train leaves at twenty-five past nine in the morning. The train journey takes fifty minutes. At what time does the train arrive at its destination?

8 Twenty-five per cent of a number is sixteen. What is the number?

9 What is half of one hundred and sixty-eight?

10 In a group of ninety-two children, thirty-seven are boys. How many are girls?

11 Ten per cent of a number is twenty-four. What is the number?

12 Write seven and a half million in figures.

13 How many thirds are there altogether in four and two thirds?

14 Kate scored fourteen out of twenty in a test. What percentage did she get?

15 Two angles in a triangle are each thirty-five degrees. What is the third angle?

16 John and Kate share some money in the ratio of three to four. John's share is sixty pounds. How much is Kate's share?

17 Multiply nine point nought four by one thousand.

18 Two hundred and ninety-five pupils out of nine hundred and seven walked to school. Estimate the percentage that walked to school.

19 How many fourteenths are there in four sevenths?

20 Look at your answer sheet. What is the largest integer x can be?

21 What is six divided by nought point two?

You will have 15 seconds to answer each of these questions.

22 Wallpaper costs five pounds ninety-nine pence a roll. How much will seven rolls cost?

23 Look at the angle on your answer sheet. Estimate the size of the angle in degrees.

24 Each side of a square is twenty-six centimetres. What is the perimeter of the square?

25 Look at the calculation on your answer sheet. What is twenty-eight multiplied by thirty-four?

26 Look at the equation on your answer sheet. If $x = 8$, what is y?

27 Seventy per cent of a number is 21. What is the number?

28 Look at the calculation on your answer sheet. What is five hundred and twenty-seven divided by seventeen?

29 A square has an area of four hundred square centimetres. What is the length of its sides?

30 Look at your answer sheet. Give the answer for this equation when x equals ten and y equals four.

31 The price of a book goes up from four pounds to four pounds forty-four pence. What is the percentage increase in the price of the book?

32 Look at the calculation on your answer sheet. Estimate an approximate answer.

32

TOTAL

Mental Test 2A

Answer sheet

1		
2	cm	
3		
4		5.9
5		
6	mm	
7		
8		
9		
10		
11		
12		
13		
14	%	
15	°	? 35° 35°
16	£	
17		9.04
18	%	
19		
20		$x < 47$
21		6 0.2
22	£	
23		?
24	cm	
25		$28 \times 68 = 1904$
26		$y = 4x - 15$
27		
28		$17 \times 31 = 527$
29	cm	
30		$(x - y)^2$
31	%	
32		$\underline{31.5 \times 18.97}$

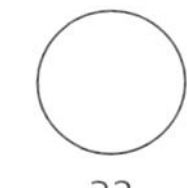

32

TOTAL

Mental Test 2B

Answer sheet

No.	Answer	
1		
2	cm	
3		
4		5.9
5		
6	mm	
7		
8		
9		
10		
11		
12		
13		
14	%	
15	°	? 35° 35°
16	£	
17		9.04
18	%	
19		
20		$x < 28$
21		6 0.2
22	£	
23	°	?
24	cm	
25		$56 \times 34 = 1904$
26		$y = 50 - 2x$
27		
28		$17 \times 31 = 527$
29	cm	
30		$(x - y)^2$
31	%	
32		$\underline{29.64 \times 42.71}$

32

TOTAL

Practice questions at Level 3

Time: 20 minutes

1 **Temperature**

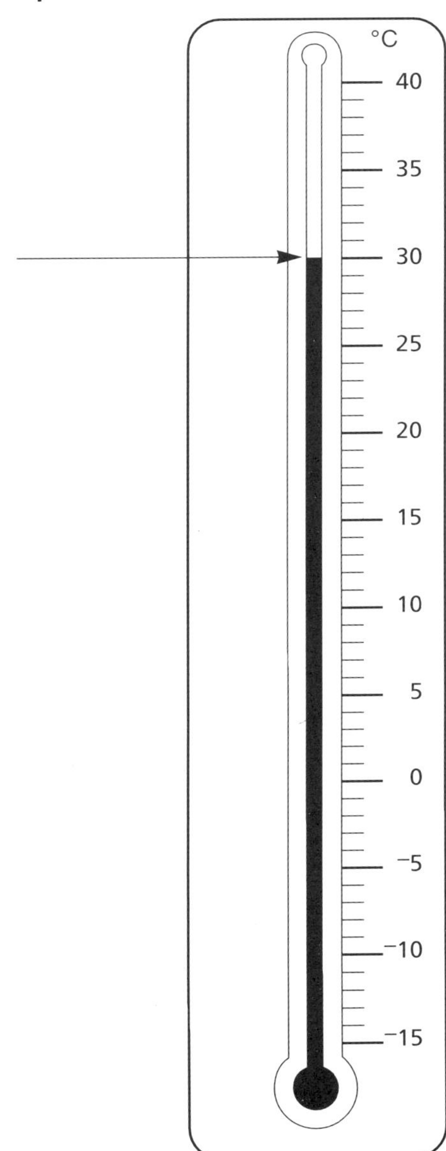

The arrow on the diagram shows where the temperature on the thermometer is **30°C**.

(a) Draw and label an arrow to show a temperature of **17°C**.

(b) Draw and label an arrow to show a temperature of −**6°C**.

(c) A temperature of **10°C** falls by **12°C**. What is the new temperature? ________ °C

(d) Put this list of temperatures in order with the coldest first.

−1°C **10°C** **−5°C** **0°C** **25°C**

____ °C ____ °C ____ °C ____ °C ____ °C

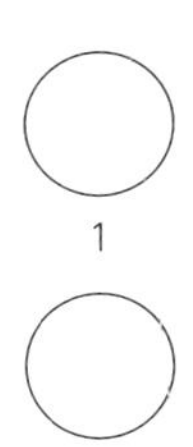

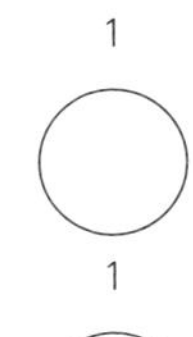

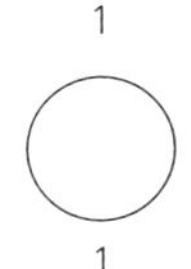

1

TOTAL

2 Shapes

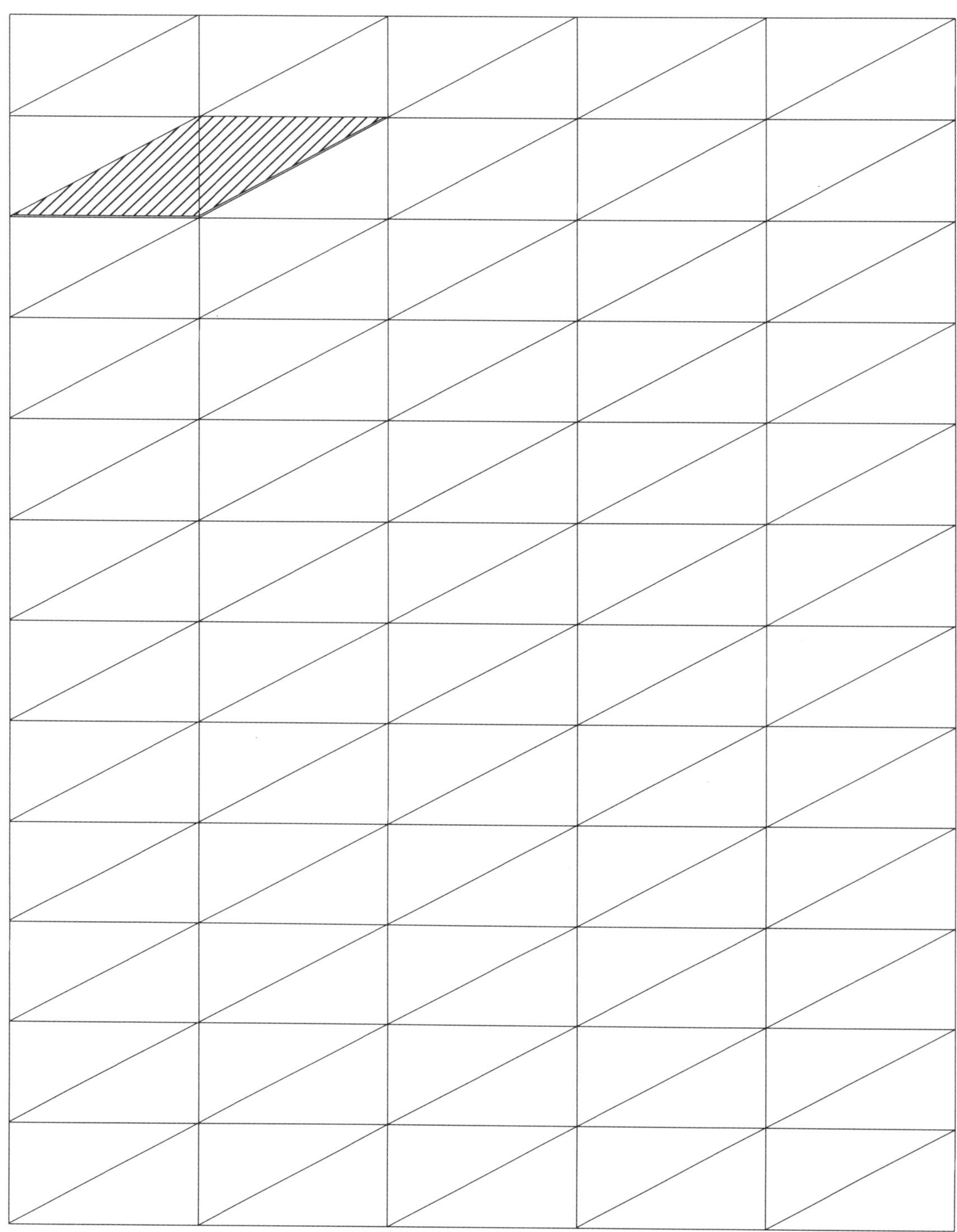

Jean has shaded **2 triangles** on her grid to make a shape with **4 sides**.

(a) Shade **2 triangles** to make a different shape with **4 sides**.

(b) Shade **2 triangles** to make another different shape with **4 sides**.

(c) Shade **4 triangles** to make a shape with **4 sides**.

(d) Shade **4 triangles** to make another different shape with **4 sides**.

(e) Shade **4 triangles** to make **1 large triangle**.

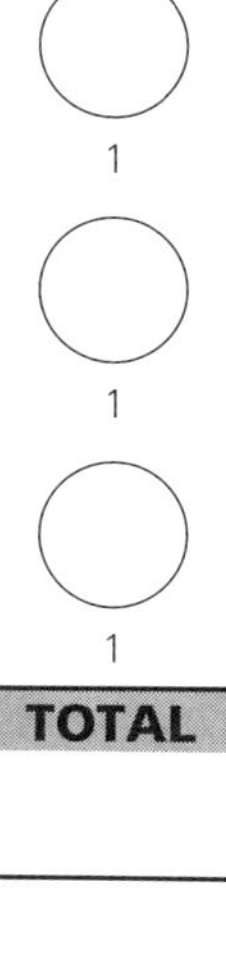

3 Bar charts

Frank's café sells pizzas. For 1 week Frank records the number of pizzas he sells each day. He has started to draw a bar chart of his results.

(a) On **Friday** Frank sold **15** pizzas. On **Saturday** he sold **8** pizzas.

Draw in the bars for **Friday** and **Saturday** on the bar chart.

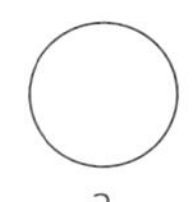

2

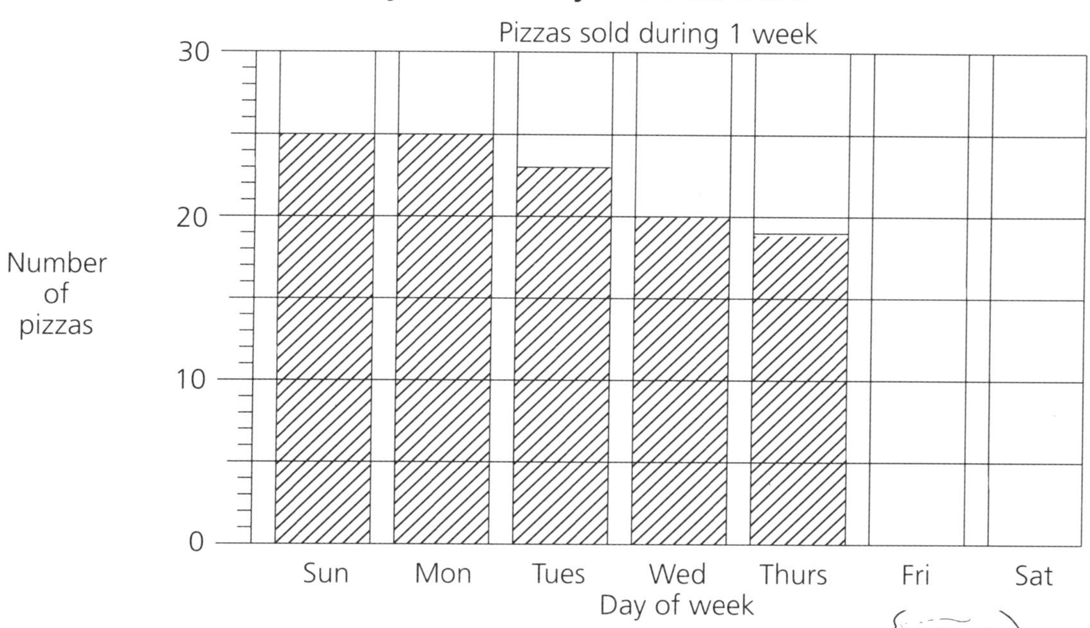

(b) How many pizzas were sold on Tuesday? ________

1

Sales of pizza were good at the start of the week but bad at the end.

FRANK'S Cafe

This graph shows the sales of meat pies at Frank's café.

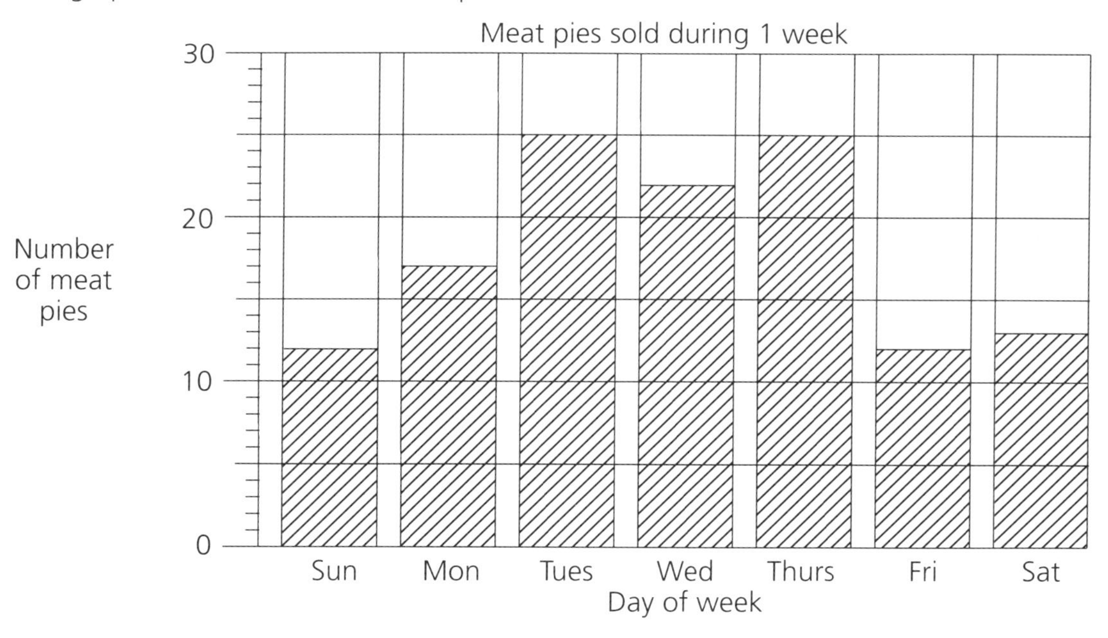

(c) Write a sentence to **describe** the **sales of meat pies**.

____

____

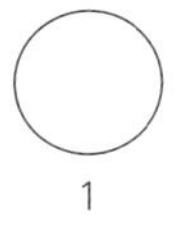

1

TOTAL

4 Distances

This table shows the distances by road between some towns. The distances are in miles.

	London	Glasgow	Edinburgh	Cardiff
Cardiff	152	392	393	- - - - -
Edinburgh	403	45	- - - - -	
Glasgow	400	- - - - -		
London	- - - - -			

(a) Which of the towns is **furthest** from **Glasgow**? ____________

1

(b) Mr Evans drives from **London** to **Edinburgh**.

How far does he drive?

________ miles

1

(c) Mr Evans then drives from **Edinburgh** to **Cardiff**.

How far is it from **Edinburgh** to **Cardiff**?

________ miles

1

(d) How far does Mr Evans drive **altogether**?

________ miles

1

TOTAL

5 Shopping

(a) Mike buys a T-shirt with a £20 note.

How much **change** should he get? £ ________

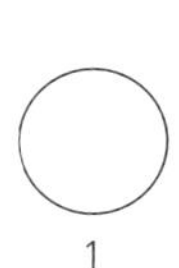

1

(b) Jackie buys:

1 T-shirt
1 pair of trainers
1 pair of shorts
1 pair of socks

How much does she pay **altogether?** £ ________

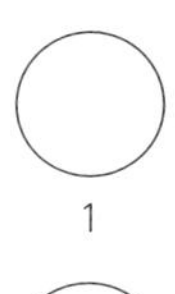

1

(c) Sara is buying T-shirts for a team. She spends **£41.40**.

How many T-shirts does she buy? ________

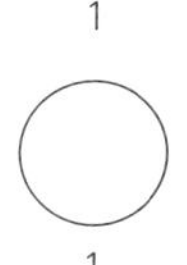

1

TOTAL

Paper 1

Test Paper 1 Levels 4–6

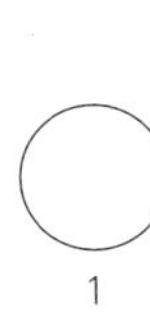

The use of calculators is not permitted in this paper. Time 1 hour

1 **Fractions and percentage**

$\frac{1}{2}$ of this shape is shaded.

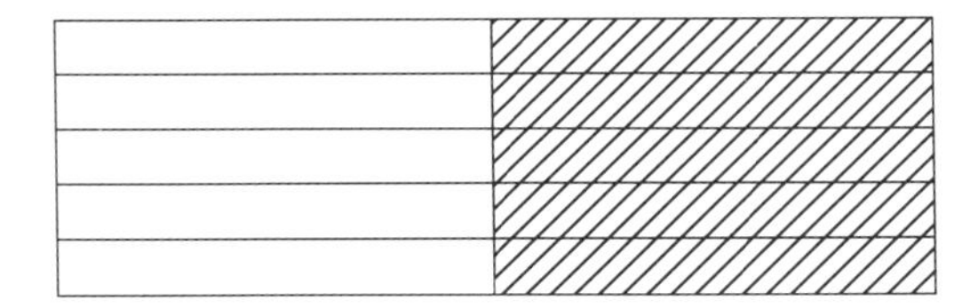

(a) Look at this shape.

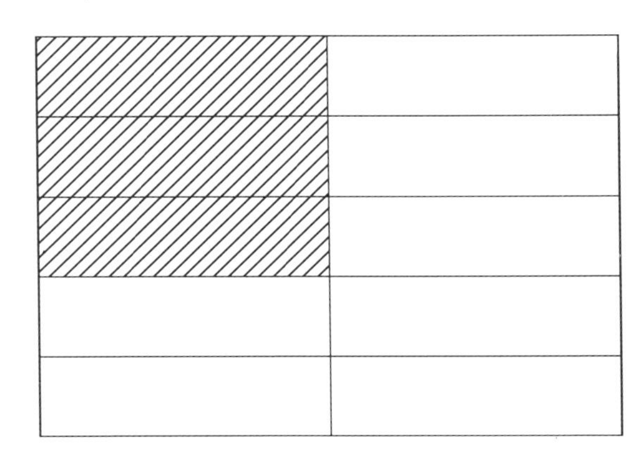

What **fraction** of the shape is shaded? ________

What **percentage** of the shape is shaded? ________ %

2

(b) Shade $\frac{3}{5}$ of this shape.

1

What percentage of the shape have you shaded? ________ %

1

TOTAL

2 Number sequences

Fill in the missing numbers in these number sequences.

(a) 7 ☐ 23 31 39 47 

1

(b) 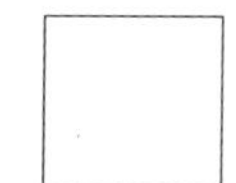 ☐ −2 1 4 7 10

1

(c) 
5.4 5.5 5.6 5.7 5.8  ☐

1

(d) Look at this number sequence and then complete the sentence.

0.31 0.32 0.33 0.34 0.35 0.36 0.37

The numbers on this number line go **up** in steps of ________

1

TOTAL

3 Symmetry

(a) Shade in **3 more squares** to make a pattern which has the dashed line as a **line of symmetry**.

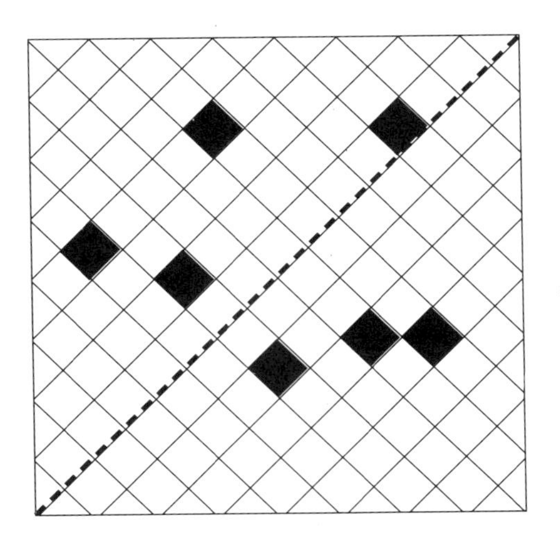

You can use tracing paper to help you.

1

(b) Shade in **3 more squares** to make a pattern which has the dashed line as a **line of symmetry**.

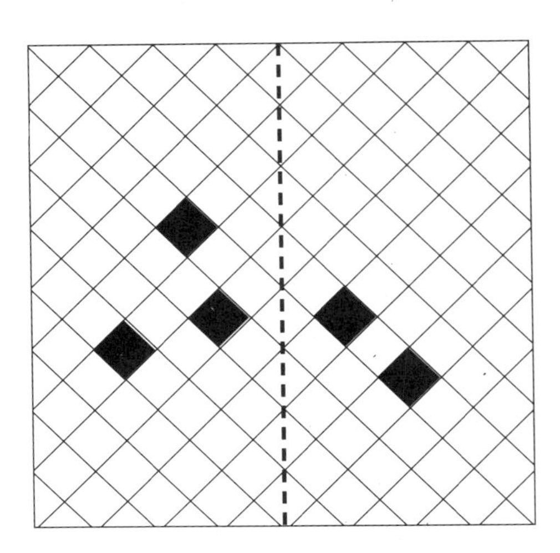

You can use tracing paper to help you.

1

(c) Shade in **6 more squares** to make a pattern which has both dashed lines as **lines of symmetry**.

You can use tracing paper to help you.

2

TOTAL

4 Number bonds

(a) Fill in the missing numbers so that the answer is **always 65**.

The first one is done for you.

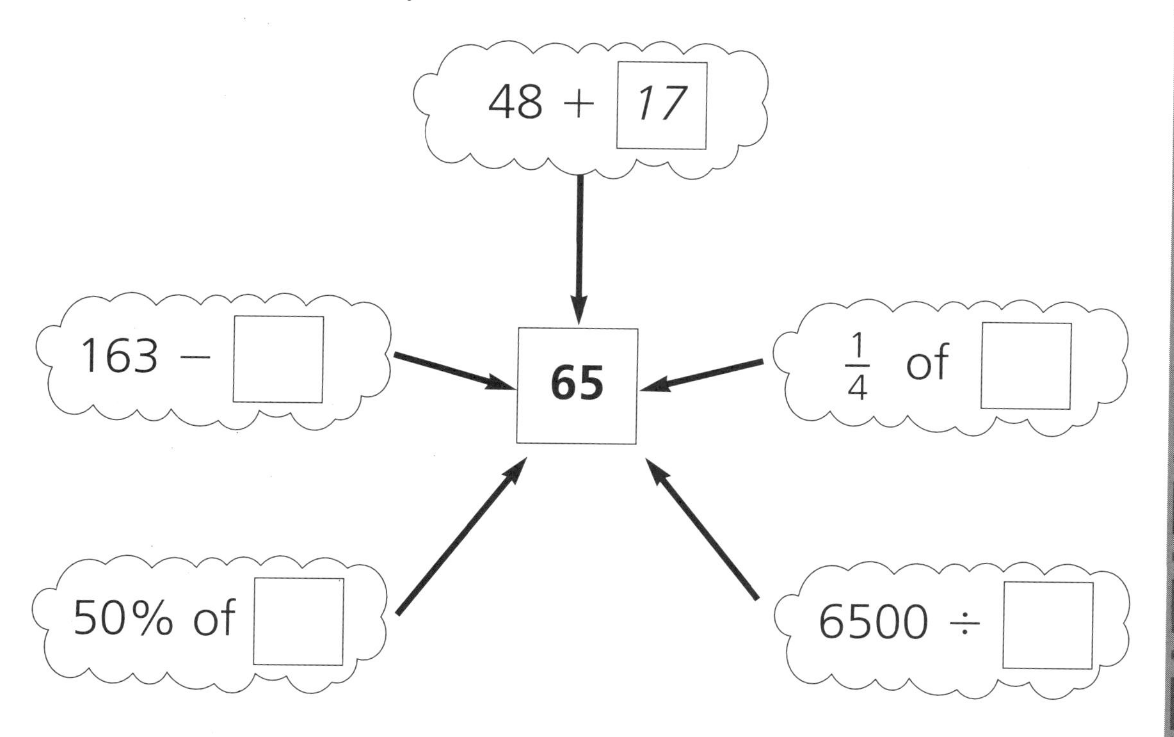

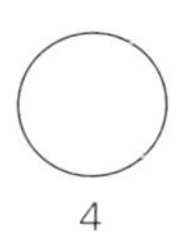

4

(b) Fill in the boxes below using any of these signs +, −, ×, ÷ to make the **answer 65**.

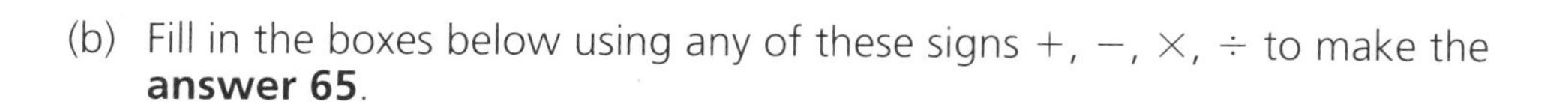

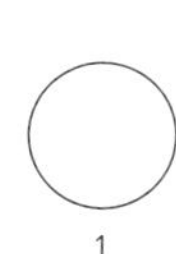

1

5 **Spinner**

(a) Patrick is going to spin this spinner.

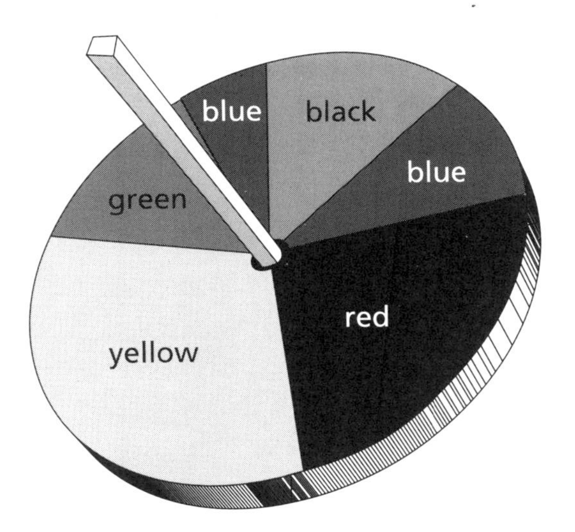

Which colour is he **most likely** to get? ________

Explain why. ____

____

1

(b) Pavneet thinks that the chance of getting **red** is about $\frac{1}{2}$. Pavneet is wrong.

Make a better estimate of the chance that she will get red. ____________

1

(c) Rob is making a spinner. He is going to use **3 colours**.
Rob wants each of the 3 colours to have **an equal chance**.
Show how Rob should divide his spinner.
Label each section with a colour: **red**, **blue** or **green**.

1

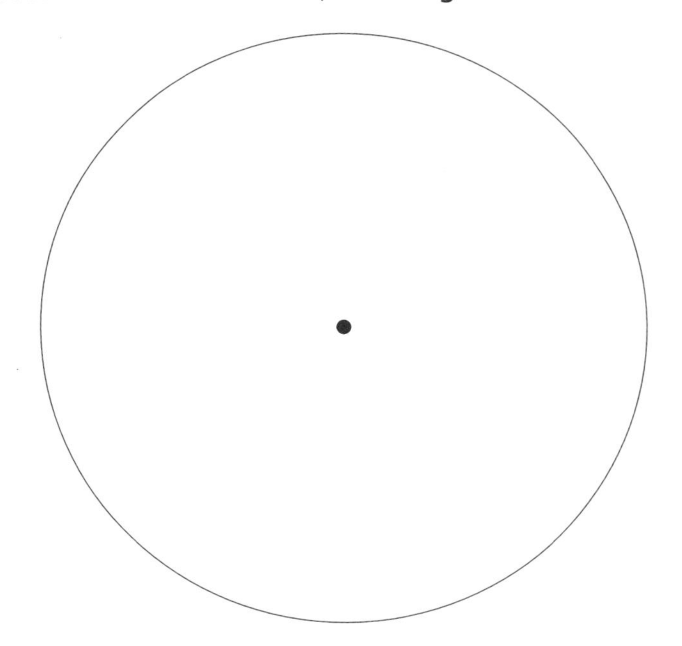

TOTAL

6 Counters

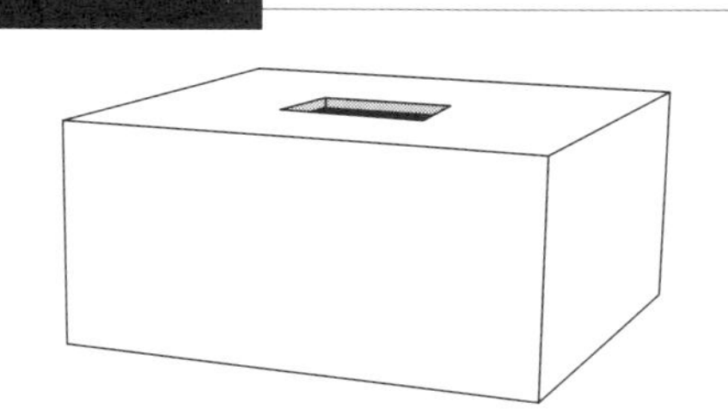

Tim has some counters in this box.

We will say that the number of counters in the box is **n**.

(a) Tim puts **three more counters** in the box.

Write a **formula** for the **total number of counters** that are now in Tim's box.

____

1

(b) Anna has **three boxes** each containing **n** number of counters.
Write a **formula** for the **total number of counters** that Anna has **altogether**.

____

1

(c) Andy has two boxes containing counters.

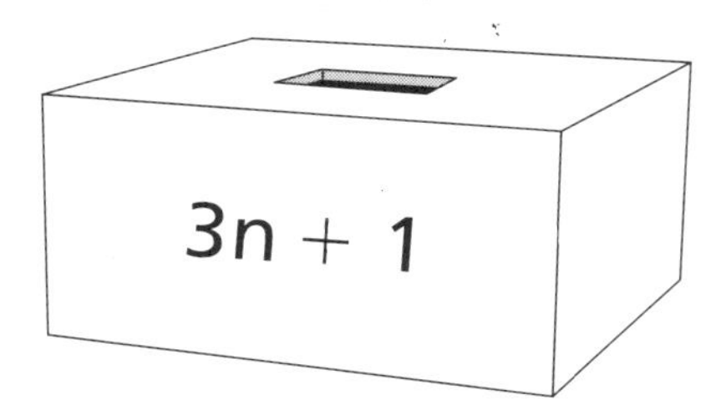

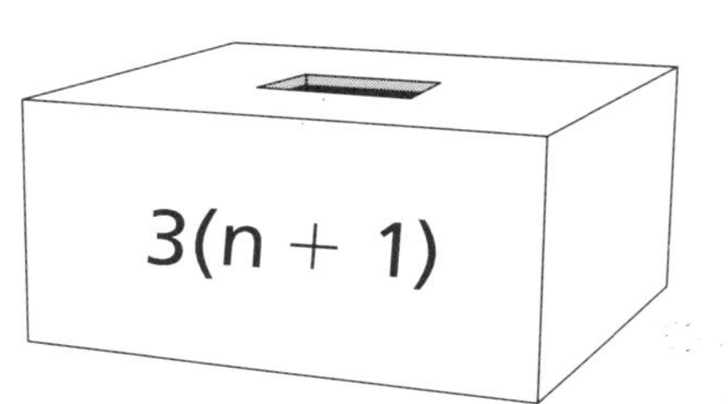

In one box Andy has **3n + 1 counters**. In the other box Andy has **3(n + 1)** counters.

Does Andy have the same number of counters in each box? __________

Explain your answer. ______________________________

____

2

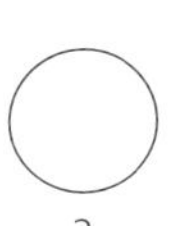

TOTAL

7 Probability

Pam and Carl each have a turn spinning this spinner.

They add their answers together to get their score.

This table shows all the possible scores.

	+	Carl 4	5	6
	4	8	9	10
Pam	**5**	9	10	11
	6	10	11	12

(a) Carl says that the **probability** of getting **11** is $\frac{2}{9}$.

Explain why Carl is right.

1

(b) Pam says that the probability of getting one of the numbers is $\frac{1}{3}$.

Which number has a **probability** of $\frac{1}{3}$?

1

(c) Label this **probability scale** with an **arrow** to show the **probability** of getting 8.

0 ——————————————— 1

1

(d) Pam and Carl each have another turn spinning the spinner. This time they multiply their numbers together to get their score.

Fill in the table to show all their possible answers.

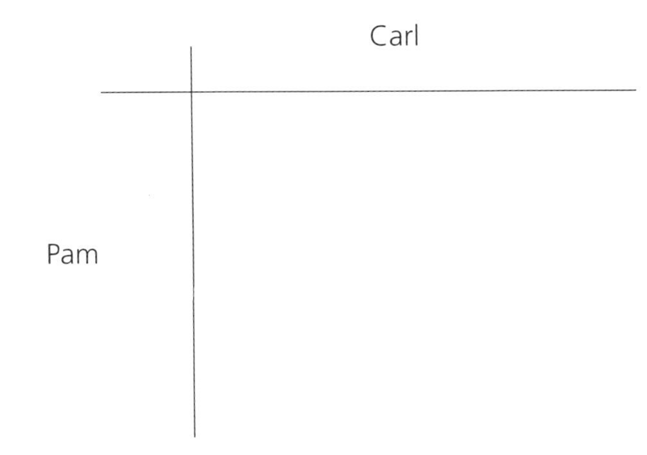

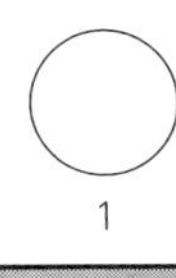

1

TOTAL

8 Snap algebra

In a card game, a player says 'Snap' if a **pair** of cards show the **same answer**.

Here are some algebra 'Snap' cards:

(a) Choose a **pair** of cards that show the **same answer**. Show your cards here.

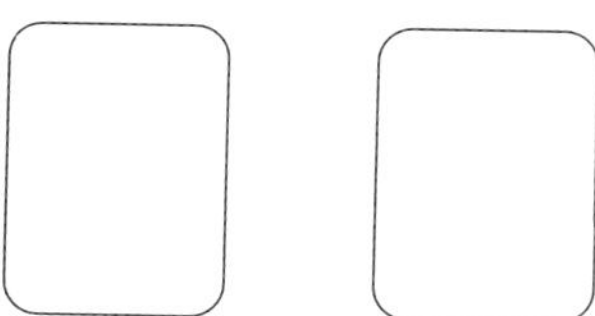

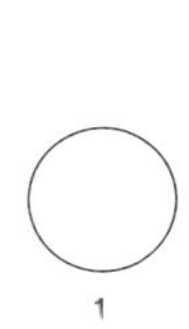

1

(b) Choose **another pair** of cards that show the **same answer**.
Show your cards here.

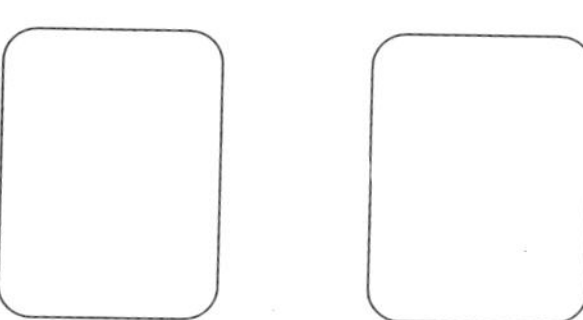

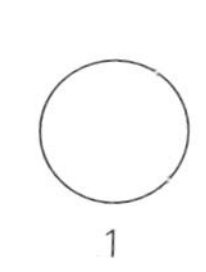

1

(c) Which card matches with this one?

Show the card here.

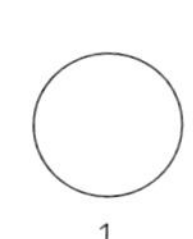

1

(d) Which card matches with this one?

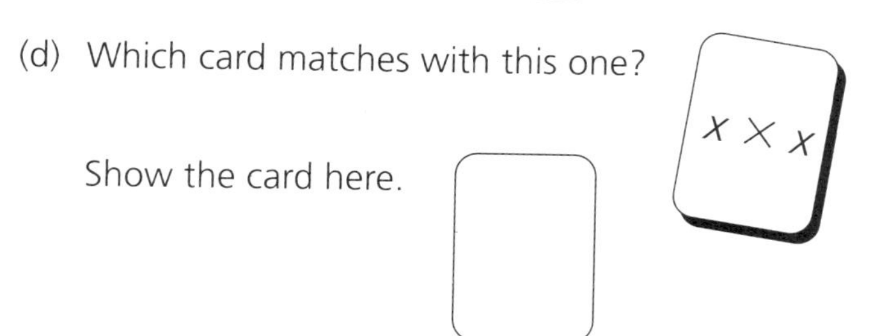

Show the card here.

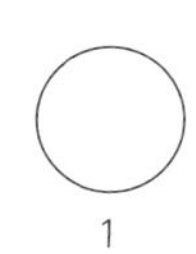

1

(e) Write a new card which shows the same answer as $x + 2x$
Show the card here.

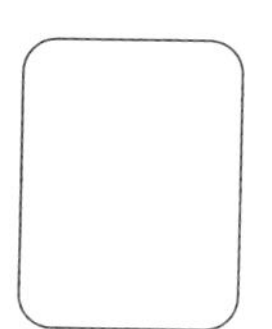

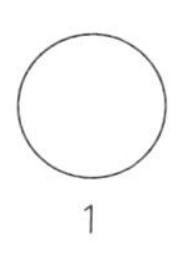

1

TOTAL

9 **Take-away**

Sam runs a shop that sells take-away meals.

(a) How much money does Sam get when he sells **34** lots of fish and chips?

£ ________

2

(b) Sam also sells chicken curry with rice.

Sam sells **17** portions of chicken curry for **£59.33**.

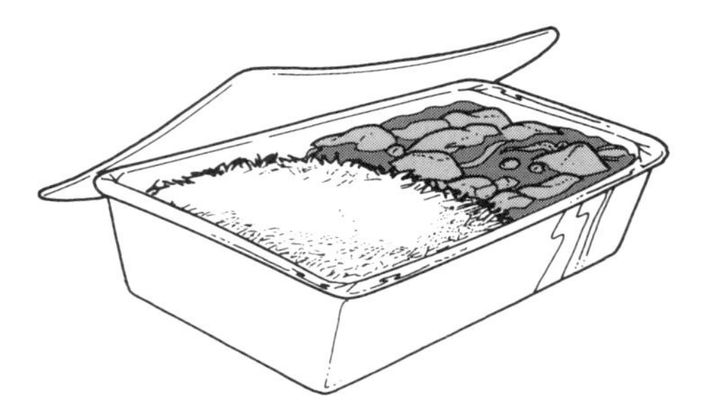

How much does Sam charge for one portion of chicken curry with rice?

£ ________

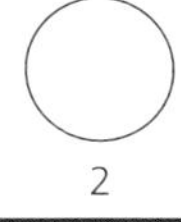

2

TOTAL

10 Number cards

Look at these number cards:

(a) Choose a card to make this sum have an answer of 1.

+ 4 + − 5 + [ ] = 1

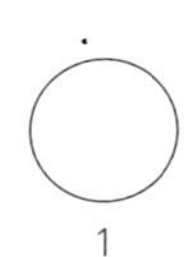

1

(b) Choose a card to give the **lowest** possible answer for this sum.

Write in both the card and the answer.

−3 + [ ] = [ ]

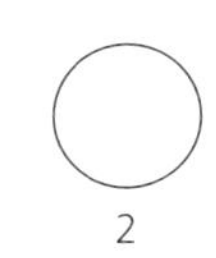

2

TOTAL

11 String

Asha has a piece of string 20 cm long. She is investigating area and perimeters of shapes.

(a) Asha makes a **square** out of her piece of string.

What is the **area** of Asha's square? ________ cm^2

1

(b) Asha makes a **rectangle** out of her piece of string. The **length** of the rectangle is **6 cm**.

What is the **width** of the rectangle? ________ cm

1

(c) Asha makes a new **rectangle** out of her string.

The **area** of the rectangle is **16 cm^2**.

Write down the length and width of the rectangle.

Length is ________ cm

Width is ________ cm

1

TOTAL

12 Polygons

Peter is using straws to make polygons.
All the straws are equal in length.

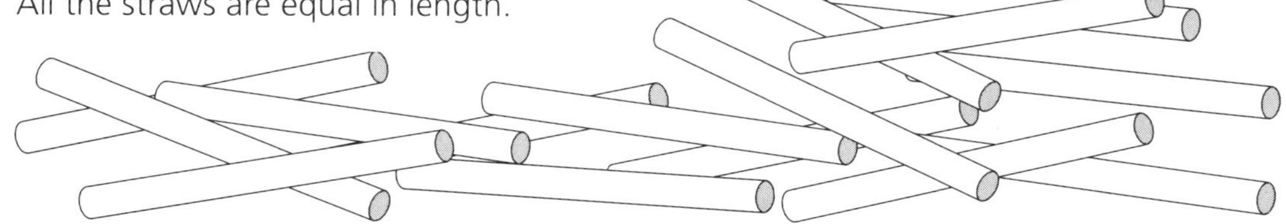

(a) Peter has made this **regular pentagon**.

Explain why angle **a** is **72°**.

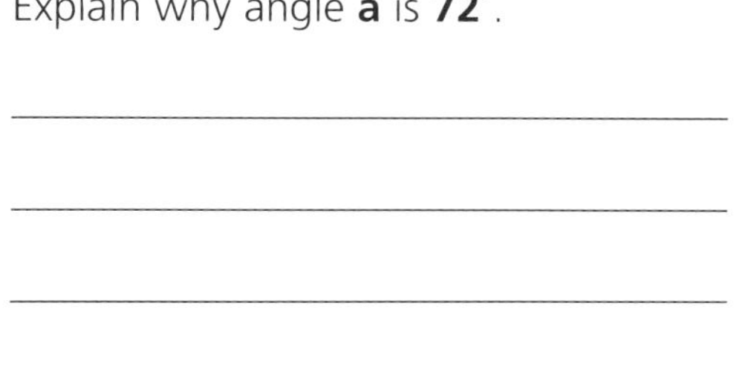

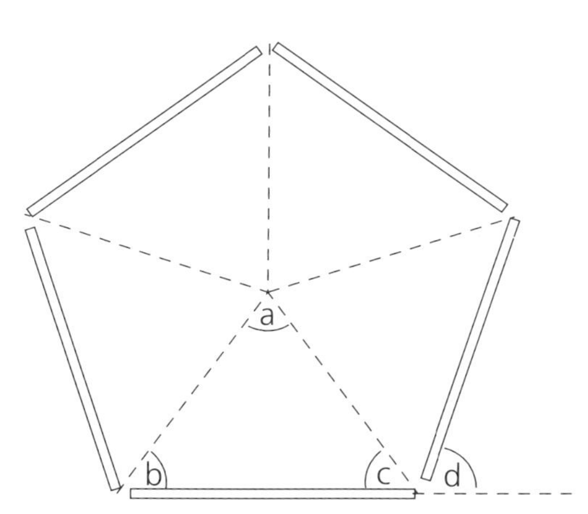

____

____

____

1

Work out the **other angles** marked.

b = ________ °

c = ________ °

d = ________ °

1

1

1

(b) Peter has made this hexagon.

Work out angle **e**.

e = ________ °

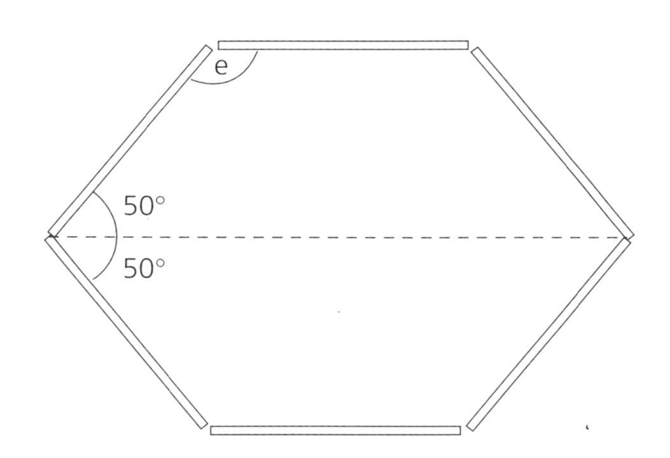

1

(c) Explain why the **hexagon** is not **regular**. ______________________________________

____

1

TOTAL

13 **Mean**

(a) Amina has these four cards:

The **mean** of these 4 numbers is **10**.

Amina takes another card.

The **mean** of the **5** cards is still **10**.

What number is on the new card? _________

1

(b) Luke starts with the same four cards as Amina.
He then takes another card.

The mean goes up by 1.

What number is on Luke's new card? _________

1

(c) Baldeep has six cards.

The **mean** of all 6 cards is **5**. The **range** of the cards is **2**.

What are the numbers on his other two cards?

_________ and _________

1

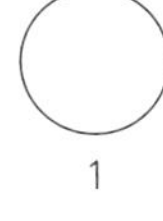

TOTAL

14 Equations

Look at this shape.

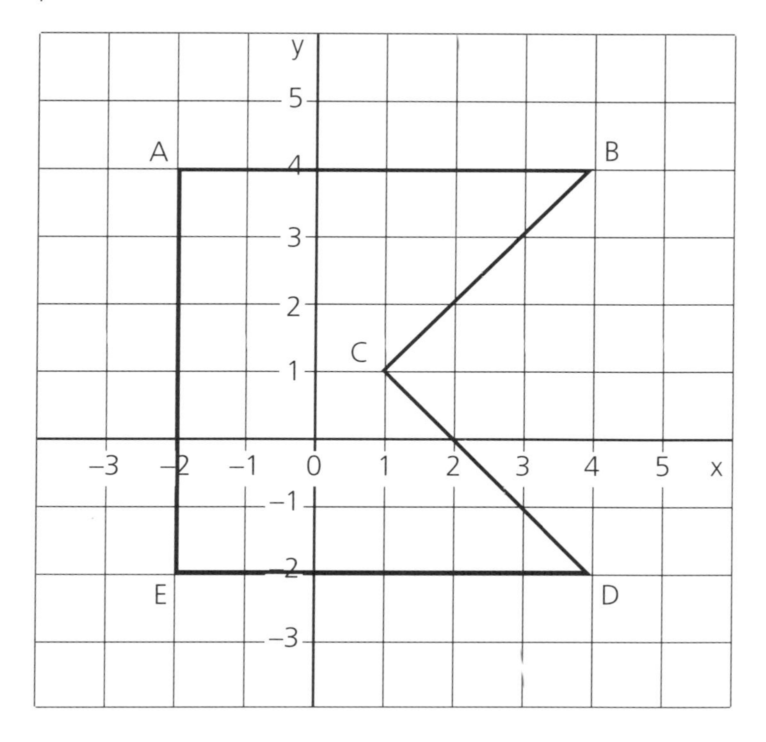

The equation of the line through A and B is ***y* = 4**.

(a) What is the equation of the line through A and E? ______________

1

(b) What 2 letters does the line y = x go through?

________ and ________

1

(c) Fill in the missing number to make this the equation for the line through C and D.

$$x + y = \square$$

1

(d) The shape has **one line of symmetry**.

Write the equation of the line of symmetry. ______________

1

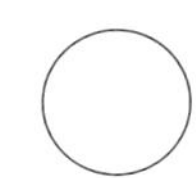
TOTAL

15 Scatter graphs

Mr Jones has a café. He wants to know if there is a relationship between what he sells and the amount of rain.
Mrs Jones works at a cinema. She thinks there is a relationship between the tickets she sells and the amount of rain.
Mr and Mrs Jones each recorded some data and then plotted these three scatter graphs.

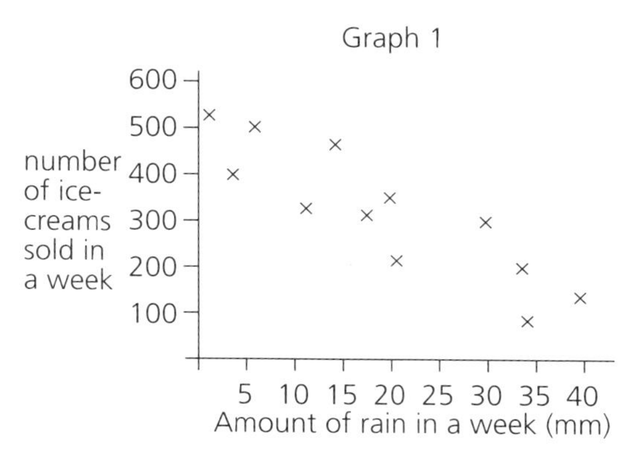

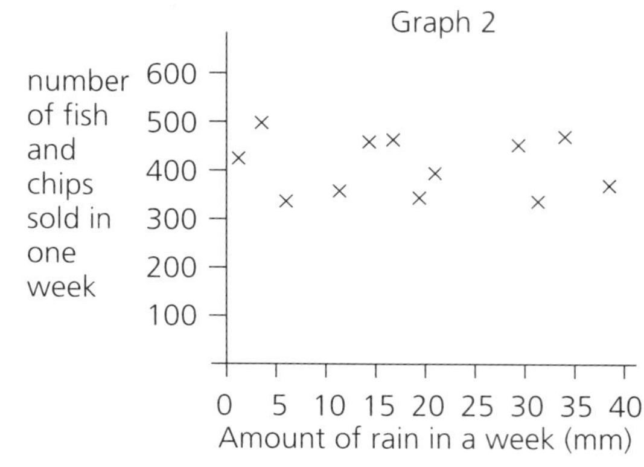

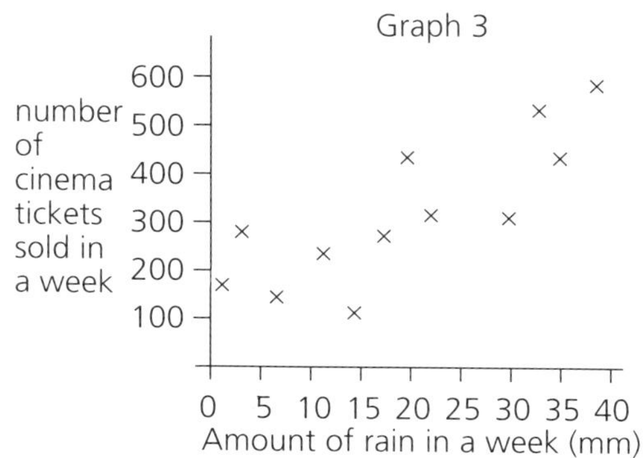

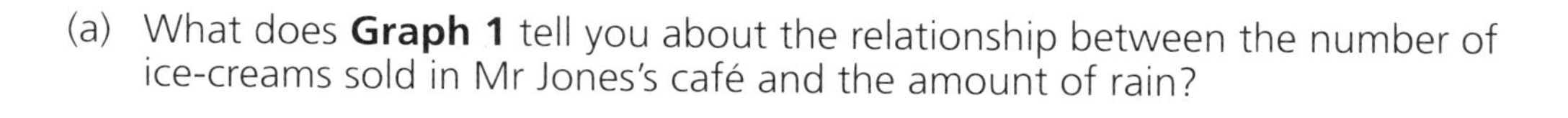

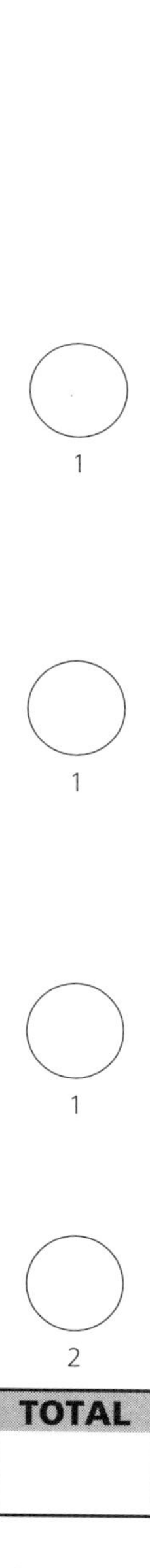

(a) What does **Graph 1** tell you about the relationship between the number of ice-creams sold in Mr Jones's café and the amount of rain?

____

____

1

(b) What does **Graph 2** tell you about the relationship between the number of fish and chips sold in Mr Jones's café and the amount of rain?

____

____

1

(c) What does **Graph 3** tell you about the relationship between the number of cinema tickets sold and the amount of rain?

____

____

1

(d) Estimate the number of ice-creams sold in a week which has a total of 25 mm of rain. ______________

Explain how you decided on your estimate. ______________

____

2

TOTAL

Paper 2

Test Paper 2 Levels 5–7

Time: 1 hour.

1 Open days

Billy's school is having 4 open days when visitors come into the school to see pupils at work.

n number of visitors came to Billy's class on Tuesday.

(a) The same number of visitors came on **Wednesday** as came on **Tuesday**.

How many visitors came on **Wednesday**?

____

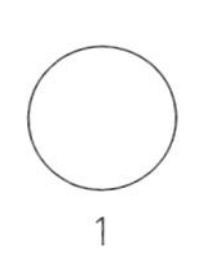

1

(b) 10 more visitors came on **Thursday** as came on **Tuesday**.

How many visitors came on **Thursday**?

____

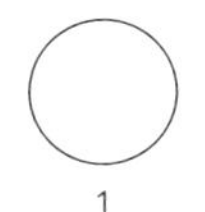

1

(c) Twice as many visitors came on **Friday** as came on **Tuesday**.
How many visitors came on **Friday**?

____

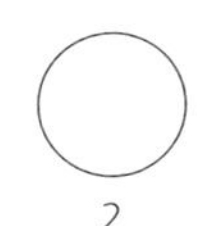

2

(d) How many visitors came **altogether**? Give your answer in the simplest way possible.

TOTAL

2 Pyramids

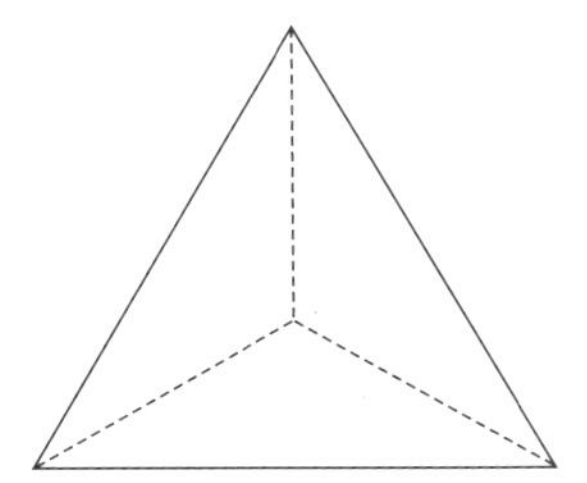

This **pyramid** has a triangular base.
The pyramid has **4 faces**.

(a) Steve has started to draw a **net** of the pyramid. He must add **one more triangle** to complete the net.

On each diagram, show a **different** place to add the extra triangle to complete the net.

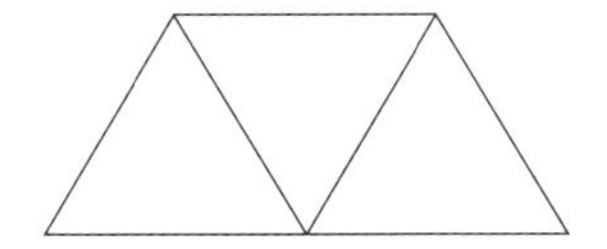

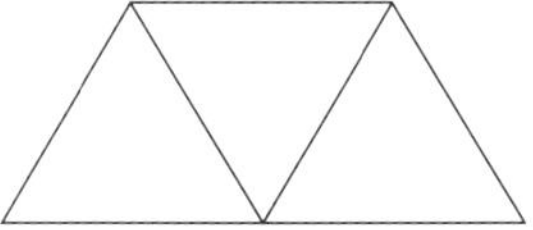

2

(b) Steve has drawn a square based pyramid. The pyramid has **four** triangular faces.

Draw **one** of the triangular faces of the net.
Draw your triangle **accurately**.

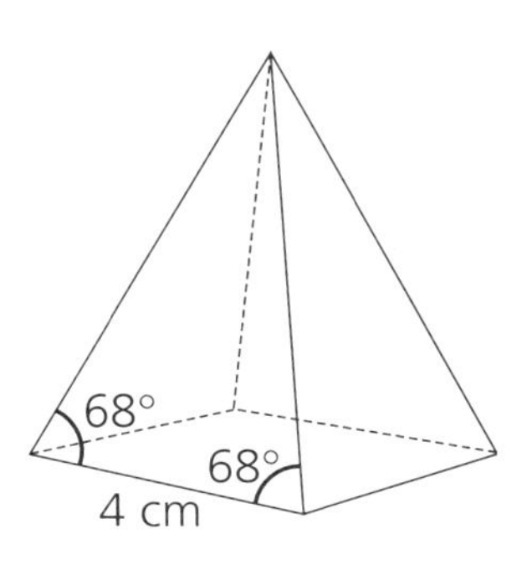

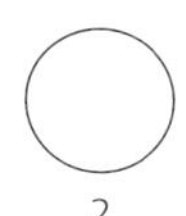

2

TOTAL

3 Graphs

Sid's Bike Hire hires out mountain bikes for children and adults. The graph shows the cost of hiring the bikes for up to 7 hours.

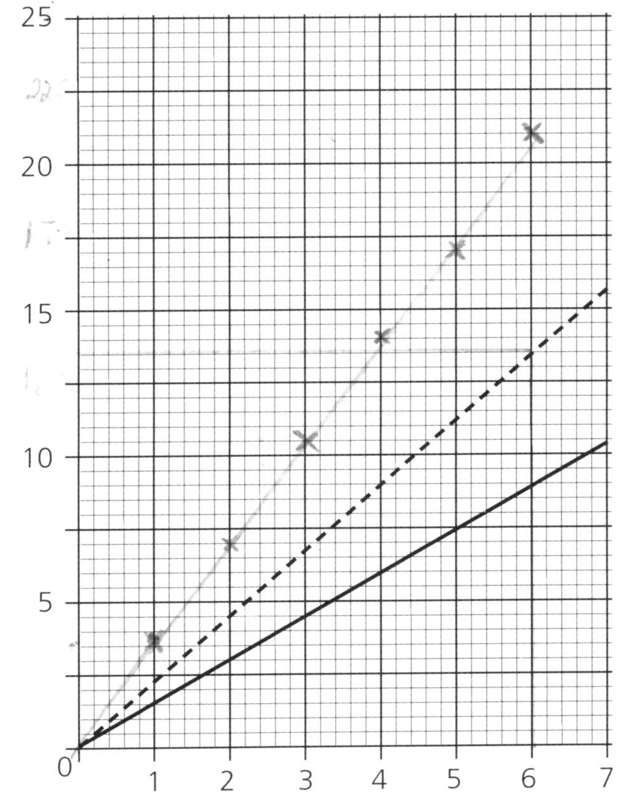

(a) Megan hires a **child's** bike for **3 hours**.

How much does this cost her? £ ________

1

(b) Lee pays **£13.50** for hiring an **adult** bike.

For how many hours does he hire the bike? ________

1

(c) Sid's Bike Hire decides to offer tandems for two people.
The hire charge for a tandem is **£3.50 an hour**.
Draw a line on the graph to show the cost of hiring a tandem.
You may want to make a table first to help you draw your line.

2

TOTAL

4 Conversions

Liz has an old recipe for potato soup.

Potato soup

2 pounds of potatoes

an onion

large lump of butter

$3\frac{1}{2}$ **pints** of chicken stock

salt and pepper

1. Peel the potatoes and peel and slice the onion. Boil them in a pan until soft.
2. Mash the potatoes with the onion.
3. Heat the mash with the butter and chicken stock.
4. Add salt and pepper to taste.

Liz only has a metric scale so she needs to change the amounts to **metric** measures. Complete the recipe so that it is a metric version. Write the **approximate amounts** and the correct **units of measure**.

Potato soup

about ________________
of potatoes

an onion

large lump of butter

about ________________
of chicken stock

salt and pepper

1. Peel the potatoes and peel and slice the onion. Boil them in a pan until soft.
2. Mash the potatoes with the onion.
3. Heat the mash with the butter and chicken stock.
4. Add salt and pepper to taste.

2

2

TOTAL

5 Special offers

David is at the supermarket. He likes to look for special offers.

(a) The packets of kiwi usually contain 10 fruit.

How many fruit will David if he gets 20% extra free?

Show your working.

1

(b) David usually pays 40p for a lettuce.

How much will he pay now?

Show your working.

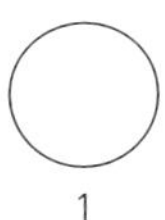

1

(c) By what percentage has the price of melons been reduced?

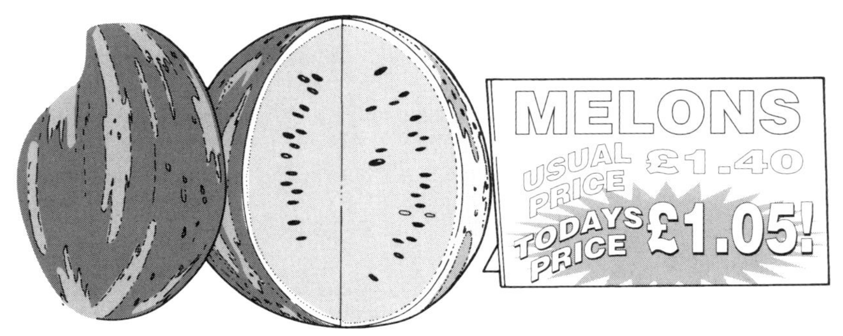

Show your working.

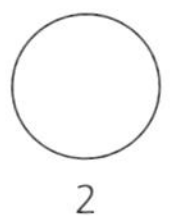

2

TOTAL

6 **Survey**

Some pupils wanted to survey people in their area to find out if they wanted a larger swimming pool. To do this they wrote a questionnaire.

(a) One question was: *How much are you prepared to pay to go swimming?*

£1 or under ☐   £1 to £1.50 ☐   £1.50 to £2 ☐   £2 to £2.50 ☐   over £2.50 ☐

Carl said:

The labels for the middle three boxes need changing.

Explain why Carl is right. ______________________________

____

1

(b) Another question was: *How often do you go swimming?*

never   sometimes   quite often   very often

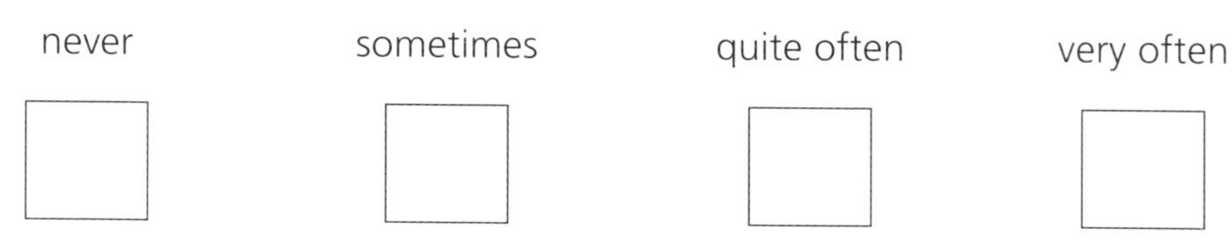

Sian said:

Some of these labels need changing.

Write new labels for any boxes that need changing.

________   ________   ________   ________

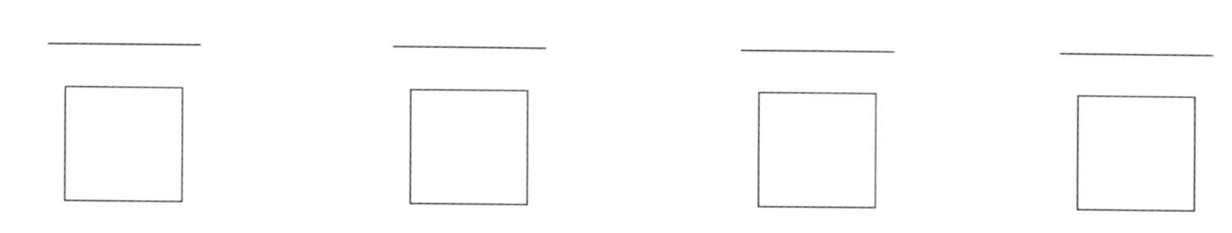

2

(c) The pupils decided to ask 50 people to fill in their questionnaire.

Richard said:

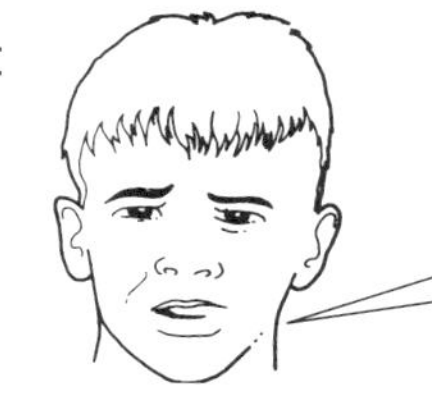

Let's ask 50 people who use the old small swimming pool.

Give **one disadvantage** of Richard's suggestion. ______________________________

____

1

TOTAL

7 Rotating

Here is a wooden block drawn on a grid.

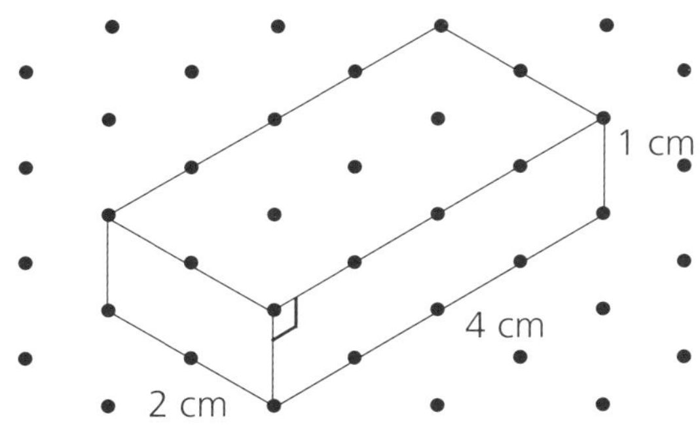

Dinah moves the block by rotating it about the **2 cm** edges. This is its new position.

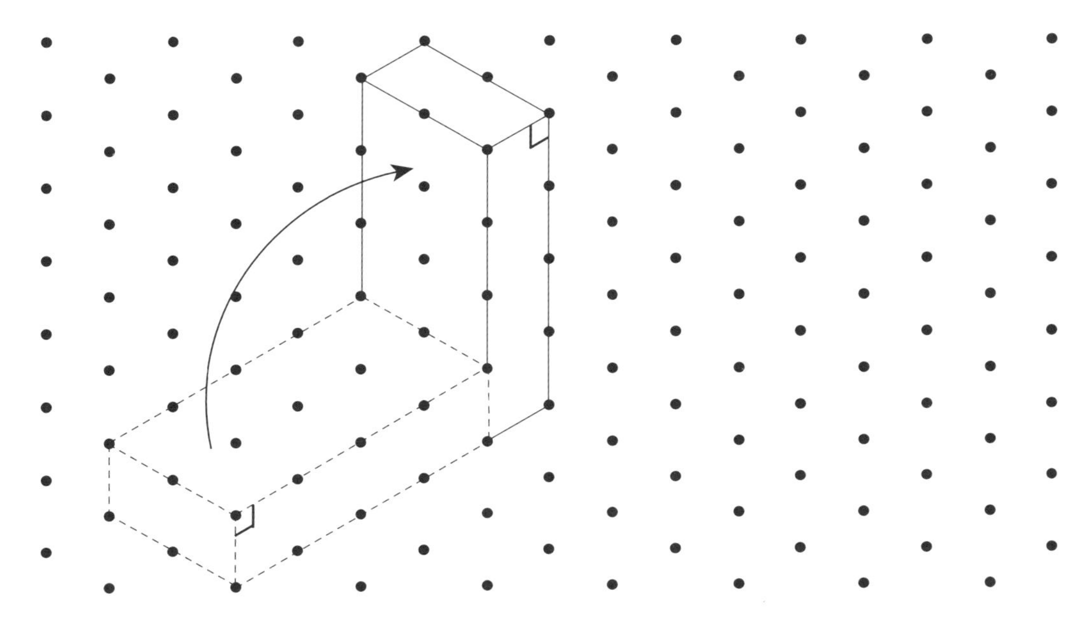

Draw the block after the next move. Remember to also draw in where the **black square** is.

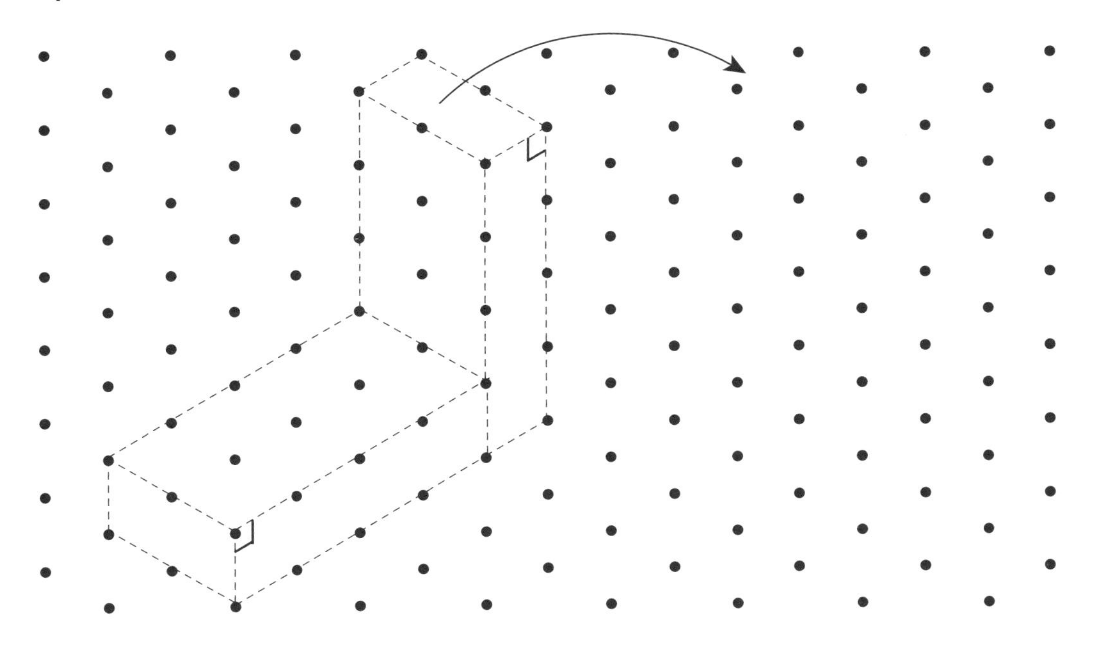

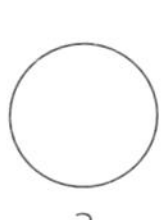

3

TOTAL

8 Survey

Balraj has done a survey to find out the favourite subject of the **120 pupils** in his year at school. He has started to draw a **pie chart** of his results.

Balraj's year (120 pupils)

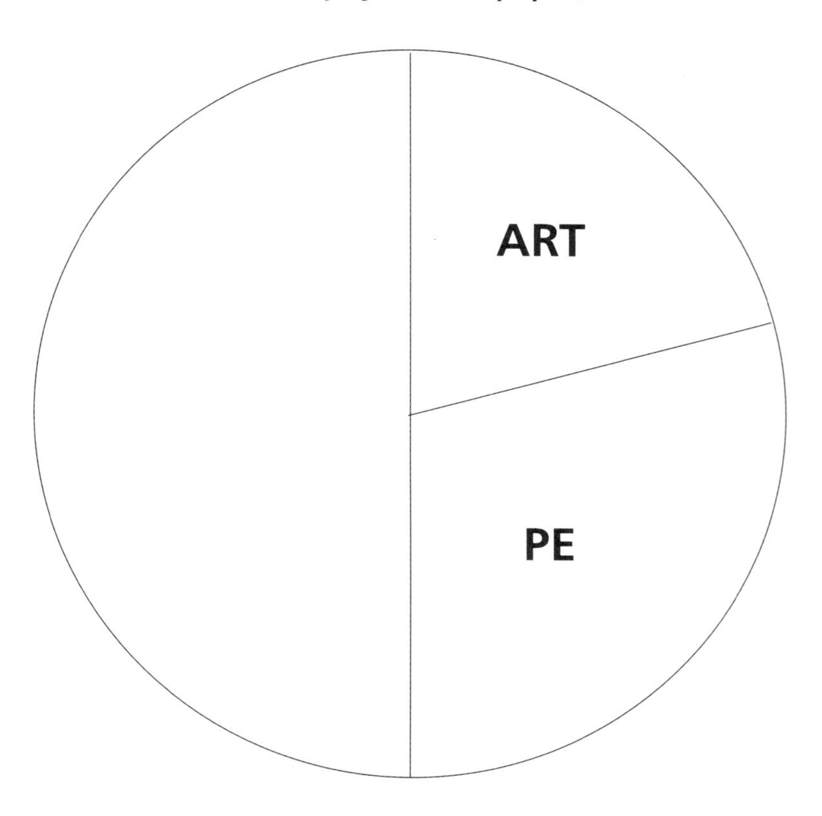

(a) The angle for the art segment of the pie chart is 75°.

How many pupils chose art? ________

1

(b) 41 pupils chose mathematics.

Show this on the pie chart as accurately as you can.

Label this part of the pie chart mathematics. Label the remaining part English.

2

TOTAL

9 Canteen

This table shows the number of pupils in a school who eat in the school canteen.

	do not eat in school canteen	eat in school canteen
girls	52	132
boys	68	102

There are **354** pupils in the school.

(a) What **percentage** of the pupils are **boys**? Show your working.

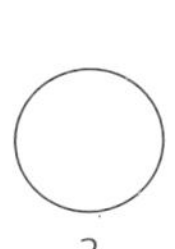

2

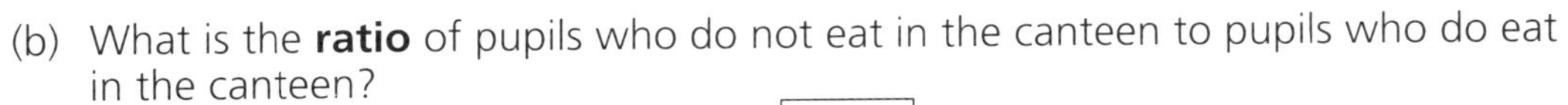

(b) What is the **ratio** of pupils who do not eat in the canteen to pupils who do eat in the canteen?

1 : ☐

Show your working.

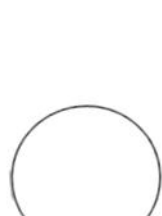

2

(c) One pupil is chosen at random from the school.

What is the **probability** that the pupil chosen is a **boy** who **does not eat** in the school canteen?

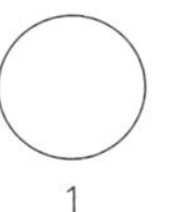

1

TOTAL

10 Rectangles

(a) Dean's rectangle has width x and length $2x + 9$.

$2x + 9$

x

Write a formula for the perimeter of Dean's rectangle. Write it as simply as possible.

____

2

Dean wants to find the width x of his rectangle. He knows that the perimeter of his rectangle is 51 cm.

Use your formula to find the value of x to give you the width of the rectangle.

Show your working.

1

(b) Kerry's rectangle has width x and length $2x + 7$.

$2x + 7$

x

Kerry wants to find the width of her rectangle. She knows that the area of her rectangle is 75 cm^2.

An equation for the area of Kerry's rectangle is: $x \times (2x + 7) = 75$

The value of x lies between two numbers which each have 1 decimal place.

Find these numbers using trial and improvement. Record the numbers you try in the table below.

x	$2x + 7$	$x \times (2x + 7)$
4	15	60

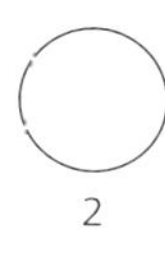

2

TOTAL

11 Triangular prisms

Here is a triangular prism.

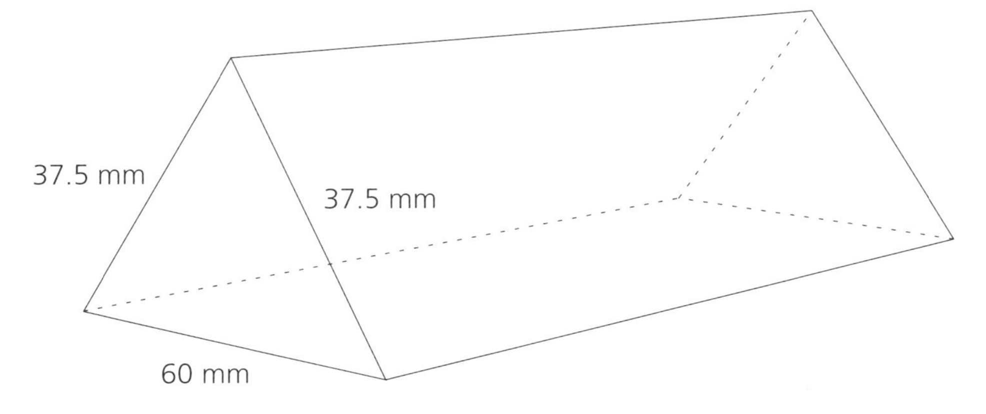

(a) Calculate the **area** of a triangular end face of the prism. Show your working.

_______________ cm^2

3

(b) The **volume** of the prism is **108 cm^3**.

Calculate the **length** of the prism.

Show your working.

_______________ cm

2

TOTAL

12 **Dots**

Mat is using triangles made up of dots to make a number pattern.

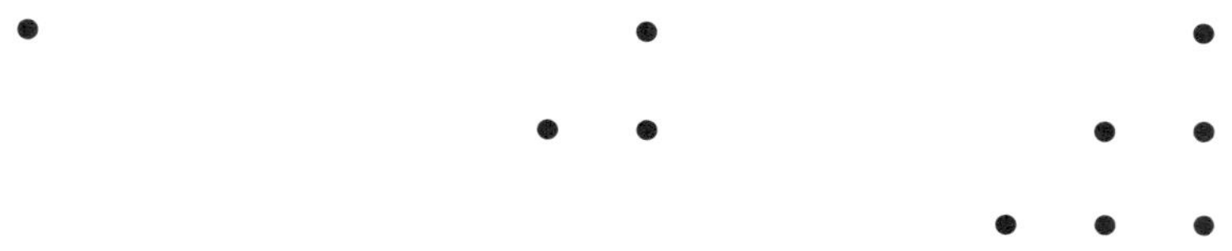

Mat's pattern:

1st term	2nd term	3rd term
$\frac{1}{2} \times 1 \times 2$	$\frac{1}{2} \times 2 \times 3$	$\frac{1}{2} \times 3 \times 4$

(a) Complete the fourth term in the pattern for this triangle.

4th term

$\frac{1}{2} \times \square \times \square$

1

(b) Mat wants to write down an expression for the **nth term** of his pattern.

Complete Mat's expression for the nth term.

nth term = $\frac{1}{2} \times \square \times \square$

2

TOTAL

13 Garden

Here is a plan diagram of Parvinda's garden.

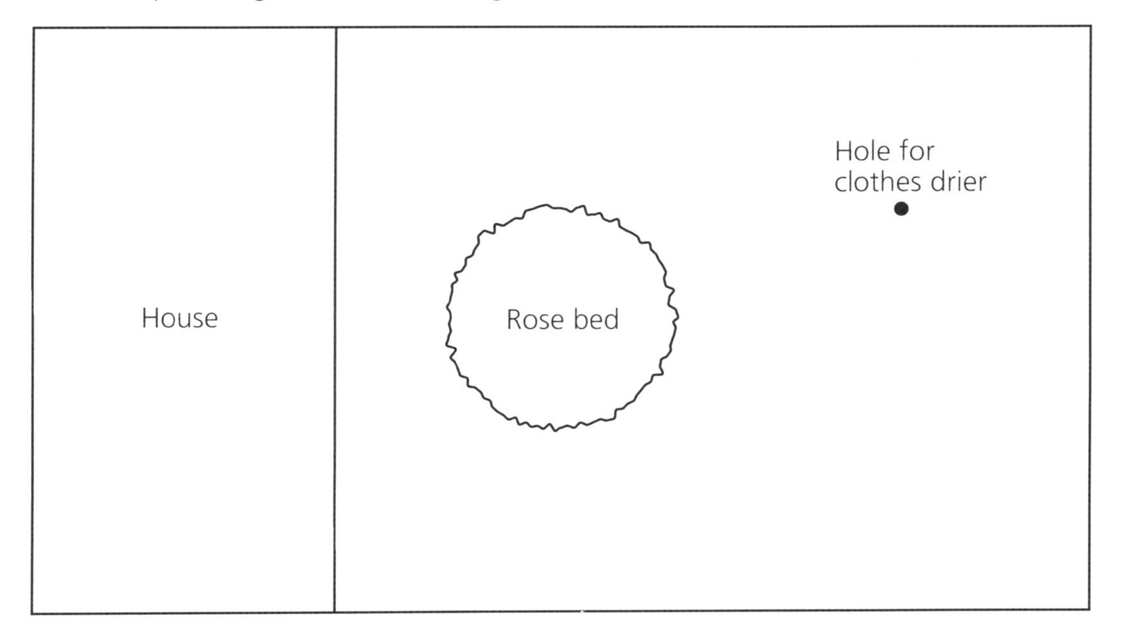

Scale: 1 cm to 1 metre

(a) Parvinda wants to plant a tree in her garden.

The tree must be **at least 6 m** from the house. The tree must be **at least 3 m** from the hole where the clothes drier is put.

Draw and shade accurately on the plan the region where Parvinda can plant the tree.

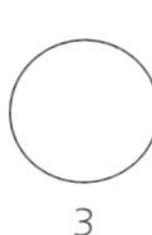

3

(b) The circular rose bed in Parvinda's garden has a radius of **1.5 m**.

Parvinda is going to make a path around the rose bed of width **0.8 m**.

Calculate the **area** of the path. Show your working.

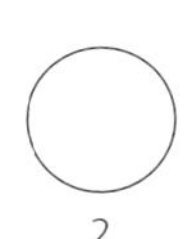

2

_____________ m^2

TOTAL

14 **Peas**

Ben has taken 50 pea pods and counted the number of peas in each pod.

His results are shown in the table.

Number of peas in a pod	Number of pods
4	3
5	8
6	11
7	13
8	8
9	6
10	1

(a) Work out the **mean** number of peas in a pod. Show your working.

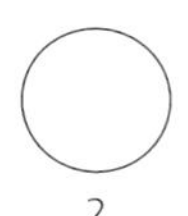

2

(b) Jill has **200** pea pods of the same variety as Ben's.

Work out the **number of pods** that Jill can expect to contain **fewer than 6** peas.

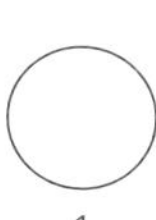

1

(c) Ben says the **probability** of getting **6 peas** in a pod is $\frac{11}{50}$.

Jill says she has **52** pods with **6** peas so that the probability of getting 6 peas in a pod is $\frac{52}{200}$.

Jill says $\frac{52}{200}$ cancels down to $\frac{13}{50}$ so Ben is wrong.

Whose answer gives a better estimate of the probability of getting 6 peas in a pod? Explain your answer.

1

TOTAL

15 Cars

(a) Alice's model car travels at a speed of 4 metres per second.

Alice times the car with a stop watch.

What time will Alice's stop watch show when the car has travelled 56 m?

__________________ sec

1

(b) Jim's model car travels 25 metres in 8 seconds.

Jim wants to know how fast this is in kilometres per hour.

Calculate the speed of Jim's car in kilometres per hour. Show your working.

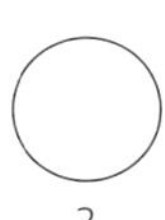

2

 kilometres per hour

TOTAL

Practice Questions at Level 8

Time: 20 minutes.

There is a formula sheet on page 6.

1 **More or less**

For each of these cards, x can be any positive number and the answers must also be positive numbers.

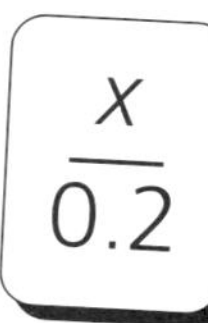

(a) When $x = 1$, which cards give the answer **1**? ______________________

2

(b) When $x = 1$, which cards give an answer **less than 1**? ______________________

1

(c) Which card will always give an answer **less than *x***? ______________________

1

(d) When x is **greater than 1**, which cards give an answer **greater than *x***?

1

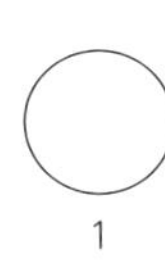

2 Formulae

One of these 2 formulae is for the **area** of the **curved surface** of a **sphere**.

The other formula gives the **volume** of a **sphere**.

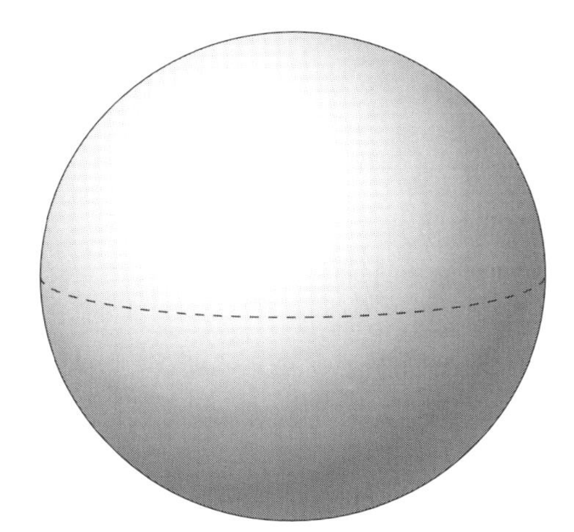

$\frac{4}{3}\pi r^3$ $\qquad$ $4\pi r^2$

(a) Ring the formula for the **volume** of a **sphere**.

Explain how you can be certain this is **not** the formula for the area of the curved surface.

____

____

1

(b) Here is another formula:

$$V = \frac{1}{3}\pi r^2 h$$

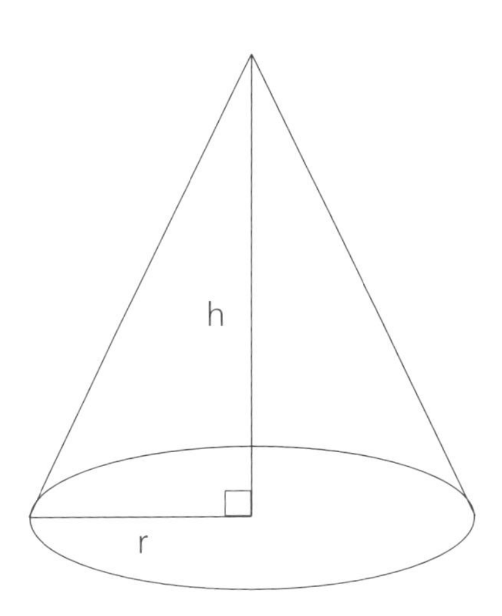

Use the formula to calculate the **radius** of the base of a cone with a **volume** of **92 cm^3** and a **height** of **5.5 cm**.

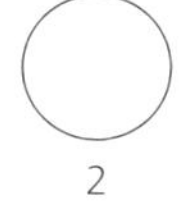

2

______________ cm

TOTAL

3 Solar system

Vicky is making a scale model of the solar system. She has found this table showing the distances of planets from the Sun. The distances are given in kilometres in standard form.

Planet	Distance from Sun (km)
Earth	1.5×10^8
Jupiter	7.78×10^8
Mars	2.28×10^8
Mercury	5.8×10^7
Pluto	5.92×10^9
Saturn	1.43×10^9
Uranus	2.87×10^9
Venus	1.08×10^8

(a) Write the distance from **Jupiter** to the Sun as a **single number**. ______________

1

(b) Which planet is **closest** to the Sun? ______________

Explain how you can tell this by looking at the numbers in the table.

____

____

____

1

(c) **Uranus** is **further** from the Sun than **Earth**.

How many times further? Show your working.

____

1

(d) In Vicky's model the **Earth** is 7.5 cm from the Sun.

What is the distance between **Uranus** and the Sun on Vicky's model?

1

TOTAL

4 Probability

A company manufacturing light bulbs tested a random sample of light bulbs from a large batch.

The company worked out the probability of a light bulb being defective as 0.02.

Andrew buys two light bulbs.

(a) Calculate the probability that **neither** bulb is defective. ________

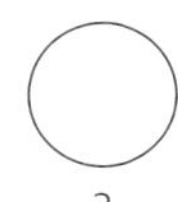

2

(b) Calculate the probability that **only one** of the light bulbs is defective. ________

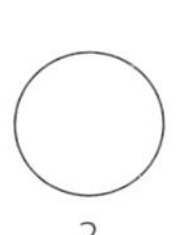

2

(c) The company found **4** defective light bulbs in the sample they tested.

Estimate the number of light bulbs that they tested. ________

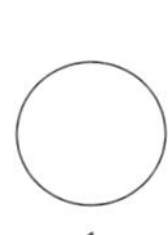

1

TOTAL

5 **Ramp**

A ramp for a wheel chair is in the shape of a triangular prism.

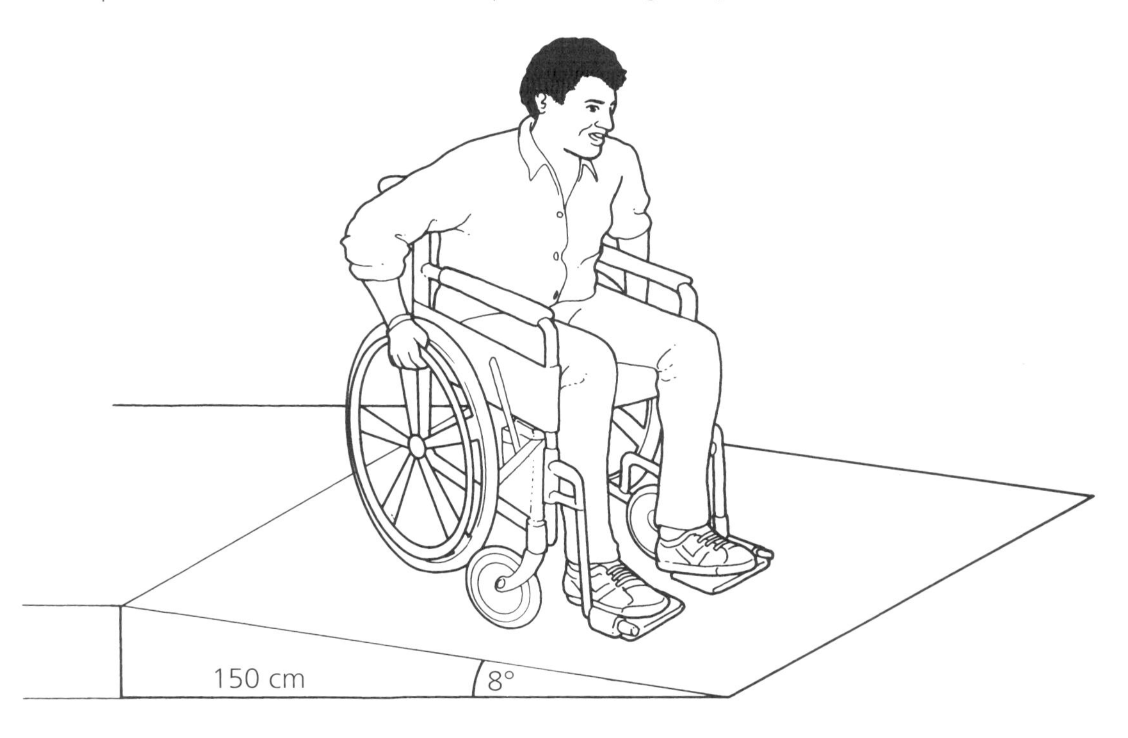

The triangular front face of the ramp is to be painted.

Find the area of the front face. Show your working.

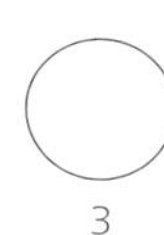

3

_______________ cm^2

TOTAL

Answers

Mental Test 1

1 30 020

2 96

3 600

4 48

5 $\frac{3}{10}$ (in 0.3, the 3 is in the tenths column)

6 850

7 8

8 250 cm ($2\frac{1}{2}$ m $\times$ 100)

9 26

10 10 (5 $\times$ 2, since there are 2 halves in one whole)

11 56% (100 $-$ 44)

12 25 m^2
(area = length $\times$ width = 5 $\times$ 5)

13 9.25 am

14 5°C (count up 8 steps from -3)

15 **A** 3.4 (scale goes up in steps of 0.2)
B 8.6

16 3 and 7 (1 and 21 divide exactly into 21 as do 3 and 7. These are the factors of 21)

17 600

18 16 (four squared written in symbols as 4^2 means 4 $\times$ 4)

19 £6.80 (try finding the difference between £8 and £14 first, then sorting out the pence)

20 34 (50% is one half. 17 is half of 34)

21 **A** 20° (accept between 15° and 25°)
B 25° (accept between 20° and 30°)

22 7

23 **A** (57), 60, 84, (93), (101)
B 80, (91), 100, 14, (107)

24 71 (do 30 + 40, then sort out the units)

25 £9.95 (5 $\times$ £2, take away 5p)

26 45 (try 60 $-$ 18 first)

27 **A** 9.5
B 6.5

28 140 (to find 35 $\times$ 4, double to get 70, then double again)

Answers

Mental Test 2

1. 1 130
2. 17 cm (170 mm ÷ 10)
3. 8
4. 590
5. 0.07 (7 goes in the hundredths column)
6. 11.5 mm (11.5 mm rounds to 12 mm)
7. 10.15 am
8. 64 (25 % is a quarter)
9. 84
10. 55 girls (try 90 − 40 first)
11. 240 (24 × 10)
12. 7 500 000
13. 14
14. 70% (multiply by 5)
15. 110° (180° − [2 × 35°])
16. £80
17. 9 040
18. 30% (about a third or 33%)
19. 8 ($\frac{8}{14}$ is the same as $\frac{4}{7}$)
20. **A** 46 ($x < 47$ means x is less than 47)
    B 27
21. 30 ($\frac{6}{0.2} = \frac{60}{2}$)
22. £41.93 (7 × £6 − 7p)
23. **A** 150° (accept between 140° and 160°)
    B 160° (accept between 150° and 170°)
24. 104 cm
    (26 cm + 26 cm + 26 cm + 26 cm)
25. 952
26. **A** 17
    B 34
27. 30
    (70% = 21, 10% = 3, so 100% = 30)
28. 31
29. 20 cm (20 cm × 20 cm = 400 cm2)
30. 36 (6^2 = 6 squared = 6 × 6)
31. 11% (44p on £4 is 11p on £1 or 11%)
32. **A** 150 $\left(\frac{30 \times 20}{4}\right)$
    B 120 $\left(\frac{30 \times 40}{10}\right)$

Answers

Practice questions at Level 3

1 **(a) and (b)**

2 marks

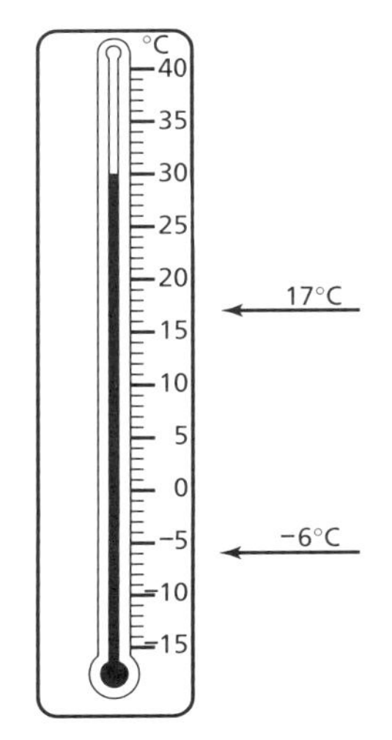

(c) −2°C *1 mark*

(d) −5°C, −1°C, 0°C, 10°C, 25°C *1 mark*

2

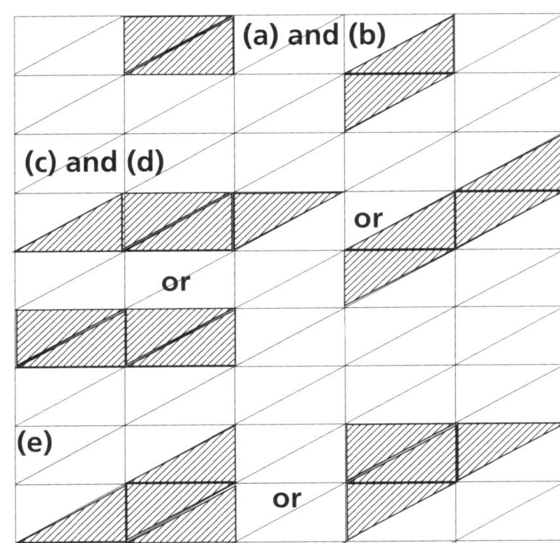

(a) and (b) *2 marks*

(c) and (d) *2 marks*

(e) *1 mark*

3 **(a)**

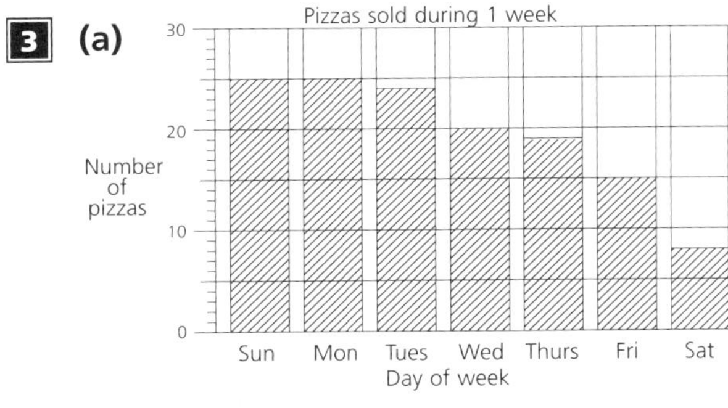

2 marks

(b) 23 pizzas *1 mark*

(c) The sales were poor at the beginning and end of the week but good in the middle. *1 mark*

4 **(a)** London *1 mark*

(b) 403 miles *1 mark*

(c) 393 miles *1 mark*

(d) 796 miles *1 mark*

5 **(a)** £16.05 *1 mark*

(b) £37.58 *1 mark*

(c) 12 *1 mark*

Answers

Paper 1

1 **(a)** $\frac{3}{10}$ *1 mark*

30% *1 mark*

(b) Any six squares shaded. *1 mark*

60% is shaded.

$\frac{3}{5} = \frac{6}{10} = 60\%$ *1 mark*

2 **(a)** 7 [15] 23 31 39 47 [55] *1 mark*

(b) [−8] [−5] −2 1 4 7 10 *1 mark*

(c) 5.4 5.5 5.6 5.7 5.8 [5.9] [6.0] *1 mark*

(d) 0.01 *1 mark*

3 **(a)** *1 mark*

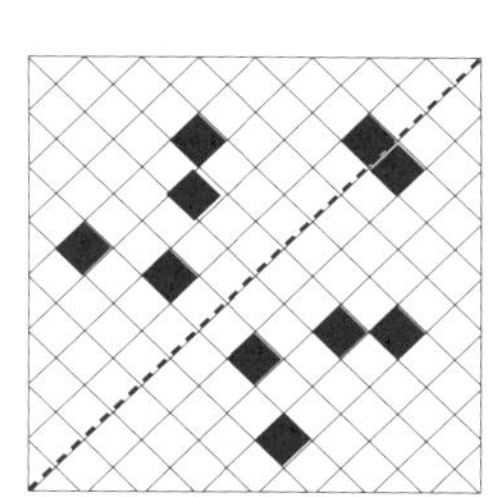

(b) *1 mark*

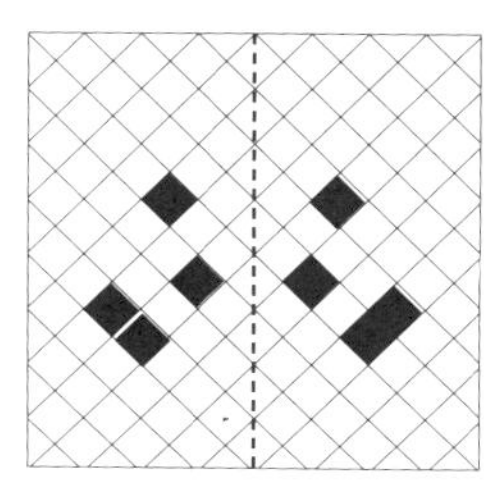

(c) *2 marks*

Hint: If your child has difficulty, try this method:

1. Put a pencil point at the centre of a shaded square. Draw a line at right angles from this point to the line of symmetry.

2. Extend the line the same distance on the other side, to find the centre of the matching square.

3. Draw and then shade the matching square. Use tracing paper as a check. Trace the dashed line and the shaded squares. The shaded squares on either side should match when the tracing paper is folded.

4 **(a)** 98 **Hint:** $163 - 65 = 98$ *1 mark*

130 **Hint:** $50\% = \frac{1}{2}$, $\frac{1}{2}$ of $130 = 65$ *1 mark*

100 **Hint:** $65 \times 100 = 6\,500$ *1 mark*

260 **Hint:** $\frac{1}{4}$ of $260 = 65$, $65 \times 4 = 260$ *1 mark*

(b) ×, + $(18 \times 2 + 29 = 65)$ *1 mark*

5 **(a)** Yellow *1 mark*

Hint: The largest slice is yellow. The answer is not blue as both blues together take up less area than the yellow slice.

(b) $\frac{1}{4}$ *1 mark*

(c) Red, blue, green each have equal slices. *1 mark*

6 **(a)** $n + 3$ *1 mark*

(b) $n + n + n = 3n$ *1 mark*

(c) Andy does not have the same number of counters in each box.

Hint: $3(n + 1) = 3 \times n + 3 \times 1 = 3n + 3$, so $3(n + 1)$ is not the same as $3n + 1$ *2 marks*

7 **(a)** There are two 11s and 9 numbers altogether. *1 mark*

(b) 10 *1 mark*

Hint: $\frac{1}{3}$ is the same as $\frac{3}{9}$.

The number 10 occurs 3 times.

(c) 0 ↓ 1 *1 mark*

Hint: The probability of getting 8 is $\frac{1}{9}$ th of the way along the scale.

(d) *1 mark*

		Carl		
	×	**4**	**5**	**6**
	4	16	20	24
Pam	**5**	20	25	30
	6	24	30	36

8 **(a) and (b)** $2x$ and $x + x$ *1 mark*

$x + 2$ and $2 + x$ *1 mark*

(c) $2 \div x$ *1 mark*

(d) x^2 *1 mark*

(e) $3x$ or $2x + x$ or $x + x + x$ *1 mark*

9 **(a)** £91.46 *2 marks*

Hint:

```
   2.69
  × 34
 1076
 8070
 91.46
```

It is useful to do estimates in this question. Estimate: £3 × 30 = £90. The answer can also be found by working out: 2.69 × 30 and 2.69 × 4 and adding.

(b) £3.49 *2 marks*

Hint:

Estimate £60 ÷ 20 = £3

```
        3.49
 17 ) 59.33
      51
       83
       68
       153
       153
```

The student may also find it helpful to work out the 17 times table first by continuous addition:

```
   17
 + 17
   34
 + 17
   51 etc
```

10 **(a)** +2 *1 mark*

Hint: +4 + −5 + +2 = 1

(b) −5 *2 marks*

Hint: −3 + −5 = −8

11 **(a)** 25 cm^2 *1 mark*

Hint: The sides of the square are each 20 cm ÷ 4 = 5 cm. The area is 5 × 5 = 25 cm^2.

(b) 4 cm *1 mark*

Hint:

Two lengths + two widths = 20 cm.
One length + one width = 10 cm.
6 cm + 4 cm = 10 cm

(c) length = 8 cm, width = 2 cm *1 mark*

Hint: length + width = 10 cm also length × width = 16 cm^2.

12 **(a)** a = 360° ÷ 5 = 72° *1 mark*

Hint: Angles at a point add up to 360°.

b = c = 54° *1 mark*

Hint: Angles of a triangle add up to 180°. b + c = 180° − a and b = c as the pentagon is regular.

d = 72° *1 mark*

Hint: Angles on a straight line add up to 180°. 54° + 54° + d = 180° so d = 180° − 108° = 72°.

(b) 130° *1 mark*

Hint: Angles between two parallel lines add up to 180°.
e = 180° − 50° = 130°.

(c) The angles of a regular hexagon are all equal. This is not the case here. *1 mark*

13 **(a)** 10 *1 mark*

(b) 15 (The new mean is 11.) *1 mark*

(c) 4 and 6 *1 mark*

Hint: The two numbers must add up to 10 to give a total of 6 × 5 = 30. The range is the largest number take away the smallest number. 4 + 6 = 10 and 6 − 4 = 2.

14 **(a)** x = −2 *1 mark*

(b) B and C *1 mark*

Hint: B is (4, 4) and C is (1,1), ie the y co-ordinate equals the x co-ordinate.)

(c) x + y = 2 *1 mark*

Hint: C is (1,1) and D (4, −2). Both of these pairs of numbers add up to 2.

(d) y = 1 *1 mark*

Hint: The line of symmetry is the horizontal line through C.

15 **(a)** The more it rains, the less ice-cream is sold. (This is called negative correlation as the crosses lie in a broad band which has a negative gradient.) *1 mark*

(b) There is no relationship as the crosses do not lie in a band. *1 mark*

(c) The more it rains, the more cinema tickets are sold. (This is called positive correlation as the crosses lie in a broad band which has a positive gradient.) *1 mark*

(d) 200 to 300 *2 marks*

Hint: The best way is to draw the line of best fit and use it to get the estimate.

Graph 1

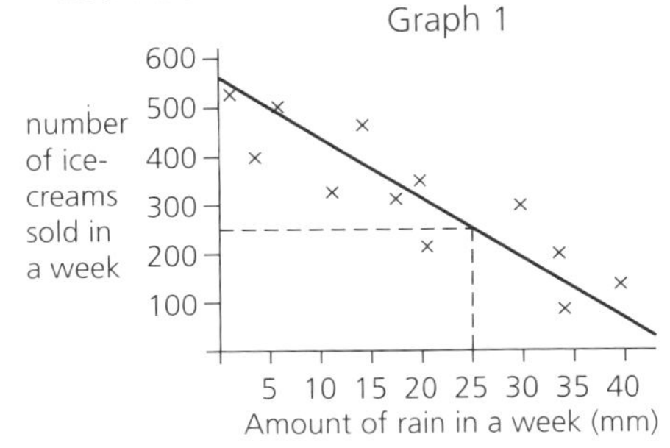

The line of best fit should have as many crosses above it as below it and should be drawn so that the cumulative distances of the crosses from the line is as little as possible.

Answers

Paper 2

1 **(a)** n *No marks*
(b) n + 10 *1 mark*
(c) 2 × n which simplifies to 2n *1 mark*
(d) 5n + 10
2 marks

2 **(a)** *2 marks*

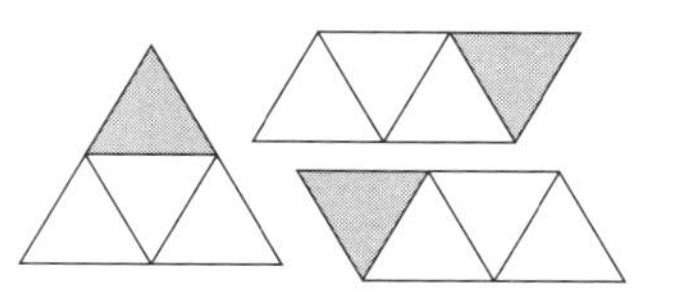

(b)

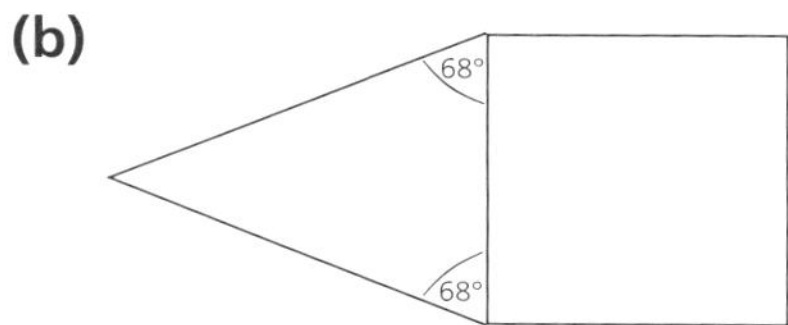

2 marks

3 **(a)** £4.50 *1 mark*
Hint: Draw a line like **a** up from 3 on the hour scale until it meets the line for child bike hire. Then follow the line across from that point to the cost scale (*y* axis) to get the answer.
(b) 6 hours *1 mark*

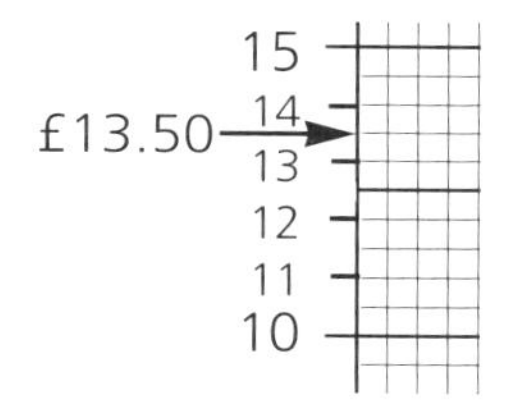

Hint: Draw a line like **b** across from £13.50 on the cost scale until it meets the line for adult bike hire. Then follow the line down from that point to the hour scale (*x* axis) to get the answer.
(c) Correct line shown on diagram. *2 marks*
Hint: Compile a table like this:

Hours	1	2	3	4	5	6	7
Cost (£)	3.50	7	10.50	14	17.50	21	24.50

You can then plot these points on your graph and join them up to make the line.

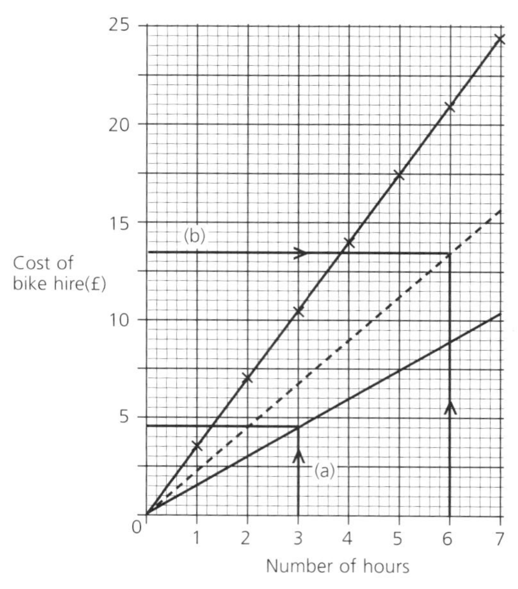

4 about 1 kilogram of potatoes *2 marks*
about 2 litres of chicken stock *2 marks*

5 **(a)** 12 *1 mark*
Hint: extra fruit = 20% of 10 = $\frac{20}{100} \times 10 = 2$.
David gets 10 + 2 = 12 fruit
(b) 34p *1 mark*
Hint: reduction = 15% of 40p = $\frac{15}{100} \times 40\text{p} = 6\text{p}$.
David pays 40p − 6p = 34p.
(c) 25% *2 marks*
Hint:
reduction = £1.40 − £1.05 = 35p
so percentage reduction = $\frac{35}{140} \times 100 = 25\%$.

6 **(a)** The boundaries of the amounts of money are not precise.
For example £1 could go in '£1 and under' or in '£1 to £1.50'. *1 mark*
(b) 'sometimes', 'quite often' and 'very often' are meaningless. Replace with precise frequencies, for example 'less than once a month', ' 1, 2 or 3 times a month', '1 or 2 times a week', 'more than twice a week'. 'never' is precise and could be left. *2 marks*
(c) The comment should be relevant. 'Nobody will tick the 'never box', or 'These people already go swimming so they may all want a bigger pool.'
1 mark

7

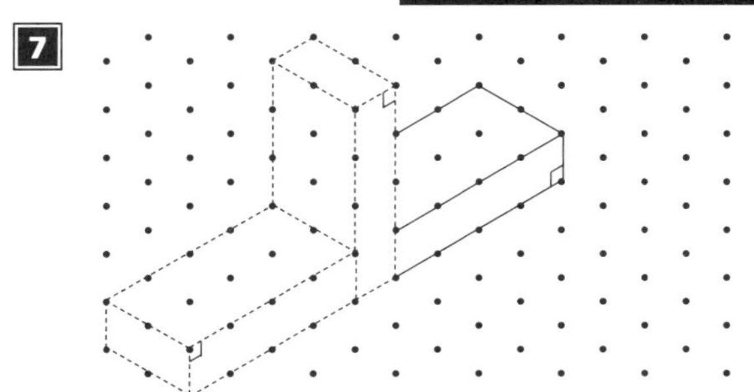

3 marks

8 **(a)** 25 pupils *1 mark*

Hint: A whole circle is 360°.
There are 120 pupils in a whole year.
One pupil is shown by
360° ÷ 120 = 3°, so 75° represents
75° ÷ 3° = 25 pupils.

(b) *2 marks*

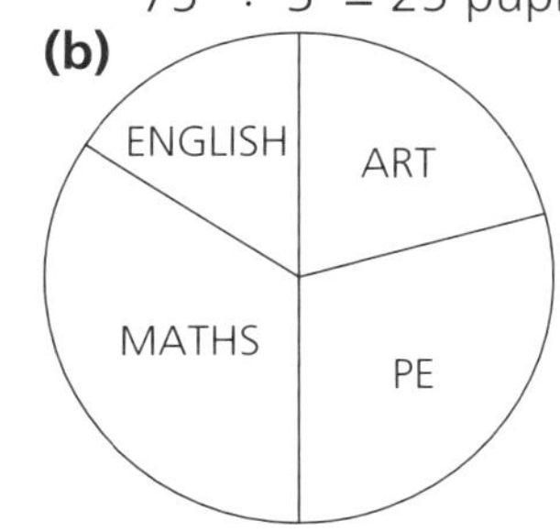

Hint:
The angle for the 41 pupils who chose mathematics should be 123° since
41 × 3° = 123°.

9 **(a)** 48% *2 marks*

Hint: There are 354 pupils in the school. Number of boys:
68 + 102 = 170.
Percentage that are boys:
$\frac{170}{354} \times 100 = 48\%$.

(b) 1 : 1.95 *2 marks*

Hint: Do not eat in canteen:
52 + 68 = 120.
Do eat in canteen:
132 + 102 = 234.
$120 : 234 = \frac{120}{120} : \frac{234}{120} = 1 : 1.95$

(c) $\frac{68}{354}$ (can be cancelled down to $\frac{34}{177}$)
1 mark

10 **(a)** Perimeter = $6x + 18$ *2 marks*

Hint: $2x + 9 + x + 2x + 9 + x = 6x + 18$

$x = 5.5$ *1 mark*

Hint: Perimeter is: $6x + 18 = 51$.
Take 18 away from both sides:
$6x + 18 - 18 = 51 - 18$ so $6x = 33$
Divide both sides by 6: $x = 5.5$

(b)

x	$2x + 7$	$x \times (2x + 7)$
4	15	60
4.5	16	72
4.6	16.2	74.52
4.7	16.4	77.08

x lies between 4.6 and 4.7 *2 marks*

11 **(a)** 675 mm² *3 marks*

Hint:
Area of triangle = $\frac{(\text{base} \times \text{height})}{2}$
so we need to find the height.

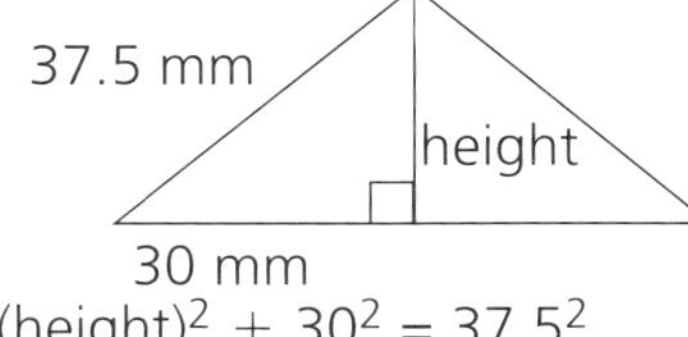

$(\text{height})^2 + 30^2 = 37.5^2$
(Pythagoras' Theorem)
$(\text{height})^2 + 900 = 1\,406.25$
$(\text{height})^2 = 1\,406.25 - 900$
height = $\sqrt{506.25}$
height = 22.5 mm
Area of triangle =
60 × 22.5 = 675 mm²

(b) 160 mm *2 marks*

Hint:
Volume = 108 cm³ =
108 × 1 000 mm³ = 108 000 mm³
Volume of prism =
area of cross-section × length
108 000= 675 × length
length = $\frac{108\,000}{675}$ = 160 mm
(The formula for finding the volume of a prism is given in the list of formulae on page 6.)

12 **(a)** 4th term = $\frac{1}{2} \times 4 \times 5$ *1 mark*

(b) nth term = $\frac{1}{2} \times n \times (n + 1)$
2 marks

13 **(a)** *3 marks*

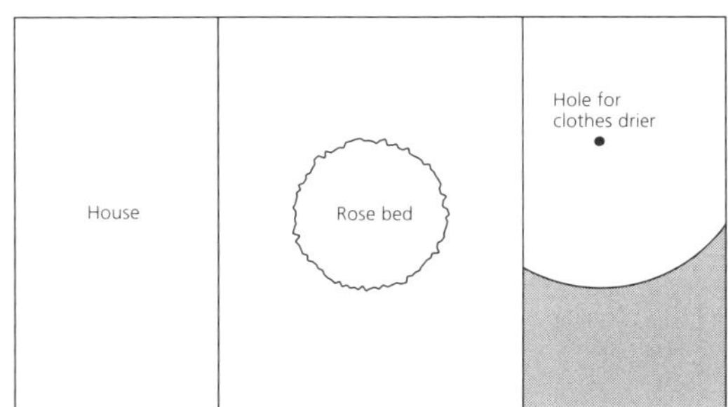

(b) 9.6 m² *2 marks*

Area of circle = πr^2 (Use π button on calculator or π = 3.14)
Radius of rose bed plus path =
1.5 m + 0.8 m = 2.3 m
Area of rose bed plus path:
π × 2.32 = 16.6 m²
Area of rose bed alone:
π × 1.52 = 7.07 m²
Area of path = 16.6 m² − 7.07 m²
= 9.6 m² correct to 1 decimal place.

14 **(a)** 6.74 *2 marks*

Hint:

Number of peas in a pod (x)	**Number of pods** (f)	$x \times f$
4	3	12
5	8	40
6	11	66
7	13	91
8	8	64
9	6	54
10	1	10
Total	50	337

Mean = $\frac{337}{50}$ = 6.74

(b) 44 pods *1 mark*

Hint: Ben has 11 out of 50 pea pods with fewer than 6 peas. This is equivalent to 44 out of 200.

(c) Both answers are estimates. Jill's is a better estimate as she has a larger sample. *1 mark*

15 **(a)** 14 secs *1 mark*

Hint: Time = $\frac{\text{Distance}}{\text{Speed}} = \frac{56}{4}$ = 14 sec

(b) 11.25 km/hr *2 marks*

Hint: Speed = $\frac{\text{Distance}}{\text{Time}} = \frac{25}{8}$

= 3.125 m/s = $\frac{(3.125 \times 60 \times 60)}{1\,000}$

= 11.25 km/h.

Answers

Practice questions at Level 8

1 **(a)** $\times 2$, $\frac{1}{x}$, $\sqrt{x}$ *2 marks if all 3 correct*
(b) $0.2x$ or $\frac{0.2}{x}$ *1 mark*
(c) $0.2x$ *1 mark*
(d) $\times 2$ and $\frac{x}{0.2}$ *1 mark*

2 **(a)** $\frac{4}{3}\pi r^3$. The formula for a volume has 3 lengths multiplied together so involves a cubed variable (eg r^3), an area only has 2 so has a squared variable (eg r^2). *1 mark*
(b) 4.0 cm *2 marks*

3 **(a)** 778 000 000 km *1 mark*
(b) Mercury *1 mark*
Hint: The smallest number in the list has the smallest power of 10.
(c) Approx. 19 times further. *1 mark*
Hint:
Divide the distance to Uranus by the distance to Earth. There is a special key on a scientific calculator that can be used for this, often marked 'EE' or 'Exp'.
(d) Approx. 143 cm (19×7.5 cm) *1 mark*

4 **(a)** 0.9604 *2 marks*
Hint: The probability that a bulb is not defective is $1 - 0.02 = 0.98$.
The probability that neither bulb is defective is $0.98 \times 0.98 = 0.9604$.
(b) 0.0392 *2 marks*
Hint: The probability that both bulbs are defective is $0.02 \times 0.02 = 0.0004$.
The probability that only one bulb is defective is $1 - 0.0004 - 0.9604 = 0.0392$.
Alternatively, the probability that the first bulb is defective and the second isn't is 0.0196. The probability that the first bulb is not defective and the second is defective is 0.0196. So the probability that only one bulb is defective is $0.0196 + 0.0196 = 0.0392$. This could be calculated using a tree diagram.

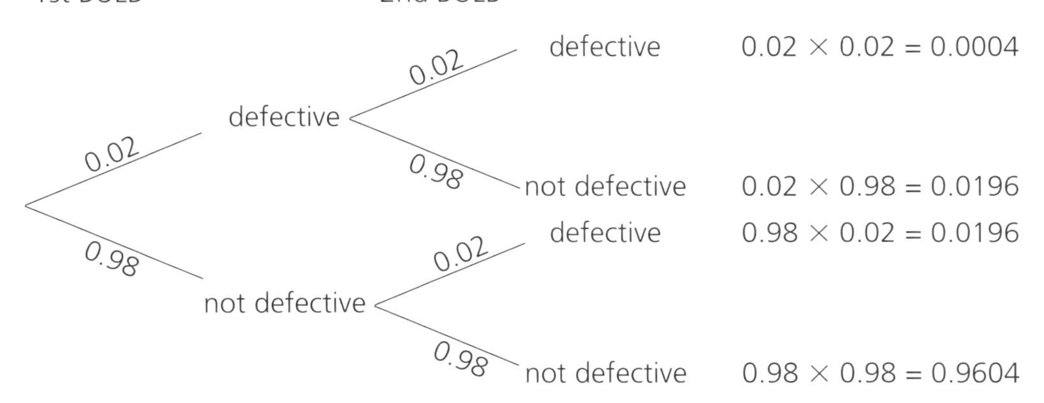

(c) 200 light bulbs *1 mark*
Hint: $\frac{4}{x} = 0.02$ so $x = 200$

5 1 421.21 cm^2 *3 marks*
Hint:
Area of triangle = base × height so we need to find the height.
height = $\tan 8^\circ \times 150$ cm = 18.95 cm
Area is $\frac{(18.95 \times 150)}{2}$ = 1 421 cm^2 to 3 significant figures.

National Curriculum Levels

Conversion of score into National Curriculum levels

1 Look at the questions which your child has got right:

Practice questions at Level 3		Level 3
Paper 1	questions 1-5	Level 4
	questions 6-10	Level 5
	questions 11-15	Level 6
Paper 2	questions 1-5	Level 5
	questions 6-10	Level 6
	questions 11-15	Level 7
Practice questions at Level 8		Level 8

2 Numerical scores can be used to give an approximate National Curriculum level.

Practice questions at Level 3: maximum marks available 20

Working at Level 3 with some areas still to be addressed	*11 – 15*
Secure at Level 6	*16 – 20*

Paper 1: maximum marks available 58

Some work at Level 4 but with many areas to be addressed	*less than 15*
Working at Level 4 and towards Level 5	*16 – 29*
Working at Level 5 and towards Level 6	*30 – 39*
Working at Level 6	*41 – 50*
Secure at Level 6	*51+*

Paper 2: maximum marks available 58

Some work at Level 5 but with many areas to be addressed	*less than 15*
Working at Level 5 and towards Level 6	*16 – 29*
Working at Level 6 and towards Level 7	*30 – 39*
Working at Level 7	*41 – 50*
Secure at Level 7	*51+*

Practice questions at Level 8: maximum marks available 20

Working at Level 8 with some areas still to be addressed	*11 – 15*
Secure at Level 8	*16 – 20*

Mental test 1
The paper is targeted at Levels 3/5.

Mental paper 2
The paper is targeted at Levels 5/6 although it contains a few questions at Levels 4 and 7.
A satisfactory mark on the paper is 16 or over.